Everyday Survival

ALSO BY LAURENCE GONZALES

The Still Point

The Hero's Apprentice

One Zero Charlie

Deep Survival

Everyday Survival

WHY SMART PEOPLE DO STUPID THINGS

Laurence Gonzales

W. W. NORTON & COMPANY

NEW YORK · LONDON

Permission granted for:
Excerpt from *The Character of Physical Law*, by Richard Feynman: MIT Press.

Printed in the United States of America
First published as a Norton paperback 2009

For information about permission to reproduce selections from this book, write to Permissions, W. W. Norton & Company, Inc., 500 Fifth Avenue, New York, NY 10110

For information about special discounts for bulk purchases, please contact W. W. Norton Special Sales at specialsales@wwnorton.com or 800-233-4830

Manufacturing by Courier Westford
Book design by Chris Welch
Production manager: Anna Oler

Library of Congress Cataloging-in-Publication Data

Gonzales, Laurence, 1947–
Everyday survival : why smart people do stupid things / Laurence Gonzales.—1st ed.
p. cm.
Includes bibliographical references (p.) and index.
ISBN 978-0-393-05838-3 (hardcover)
1. Intellect. 2. Stupidity. 3. Errors. 4. Life skills. I. Title.
BF431.G6255 2008
155.9—dc22

2008021386

ISBN 978-0-393-33706-8 pbk.

W. W. Norton & Company, Inc.
500 Fifth Avenue, New York, N.Y. 10110
www.wwnorton.com

W. W. Norton & Company Ltd.
Castle House, 75/76 Wells Street, London W1T 3QT

1 2 3 4 5 6 7 8 9 0

For Debbie

CONTENTS

Prologue 11

One The Untied Knot 19

Two Training Scars 33

Three When Modern People Face Ancient Hazards 46

Four The Sign of the Hand 57

Five Groupness 77

Six The Corporate Emotional System 97

Seven The Teachings of Don Juan 114

Eight Pleonexia 134

Nine "The Earth Is Rotting" 151

Ten The Cosmic Cheat Sheet 182

Eleven The 10,000-Watt Lightbulb 201

Twelve The Inside–Outside Problem 211

Thirteen Why We Care 220

Fourteen The Climax Shape 233

Fifteen The Guest Star 241

Sixteen Land's End 254

Epilogue 259

Select Bibliography 267

Acknowledgments 271

Index 273

The discovery of iron brought grief to men.

—*Herodotus*

PROLOGUE

When I was finishing this book, I went to stay in a house on the Outer Banks in North Carolina so that I could work far away from the distractions of daily life. The house was on the dunes above the beach, and I could sit and write and listen to the surf as it thundered beneath my window, while the constant wind blew the tops off the waves. Out on the deck I'd watch the pelicans, big and prehistoric-looking, wheel around their circuit from south to north and back again. In the angled light of afternoon, pods of dolphins leapt and dove, and children played in the waves, while I fretted about the rip currents carrying them out to sea.

When I wasn't working, I took long walks on the beach or up and down the towering dunes just to the north where the hang gliders launched not far from where the Wright brothers first took flight. Jockey's Ridge State Park has the tallest sand dunes in the eastern United States, some of them 100 feet high. But many were more modest hills, and in places they crowded together, creating networks of shadowed pathways that led into wooded areas or out onto great tumbling expanses of sea grass and live oak and even prickly pear cactus. Catbirds called from the loblolly pines, and lizards left twisted trails that looked like an ancient language scratched in the sand. More than a few times, as I hiked, contemplating some of the deeper questions in this book, I realized that I had no idea where I was. I was, in fact, lost. But the entire park is only a mile long and three-fourths of a mile wide, wedged between U.S. Highway 158 and Roanoke Sound. I knew I could find my way out.

Nevertheless, I had a hunch that this was a good place for people to find their own kind of trouble. And trouble is my specialty—studying it, writing about it, trying to understand how it happens. So I stopped a park ranger on my way out one day and asked him if people ever had to be rescued there. He just laughed. "About twenty nights a year I'm out here looking for someone," he said. Dobo Cox was his name, and he was a broad man with a low center of gravity, all muscle. He would be hard to knock over. In his midthirties, he had a great bald head and a naturally infectious smile behind wrap-around sunglasses. On his belt, crowded in among all the other paraphernalia—collapsible baton, mace, handcuffs—he wore a Glock .45-caliber pistol, the darker side of his profession. His job was trouble, too, and he appeared to have taken every precaution to keep it from getting the better of him.

Cox went on to describe the astounding failures of mind that he had witnessed as a ranger on this tiny spit of land. He told me about a group of people he'd found on top of the tallest dune in the park. They had noticed that they could see the Atlantic Ocean to the east of the park and asked him if the water to the west was the Pacific Ocean. He said he had rescued numerous people who'd become lost and were unable to think through their panic and realize that they were only a short walk from a heavily traveled road (you can hear the cars). He told me that I'd be surprised at the number of people who asked him how far it was between the mile markers on that road. Cox seemed to take it all with good humor, but he also seemed a bit in despair at the human condition. He said people come here suffering from what he called "a vacation state of mind, where all the old rules are suspended." Referring to the numerous injuries, mostly minor, that occur on the big dune, he said, "Yes, gravity still does apply here, even when you're on vacation." He said people ask him things like which side of the road to drive on. One visitor complained bitterly because the dunes were made of sand, which made them hard to climb.

I was particularly interested in talking to Cox, because I had just come from giving a talk at the Santa Fe Institute. The talk was called "Intelligent Mistakes: When Smart People Do Stupid Things," and it was a brief preview of the first part of this book. It addressed behaviors—some of them my own—that were not unlike the ones Cox was describing to me on that summer day, filled with wind and sun, on the eastern seaboard.

The Santa Fe Institute (SFI, as it's called) is one of the most respected research institutions in the nation and is well known around the world. It was formed by a group of scientists, economists, and mathematicians who believe that specialization in their disciplines, while it serves a purpose, is also limiting and has perhaps gone too far. Many of them, in fact, have done serious work in one field, only to switch to something seemingly unrelated. Murray Gell-Mann, for example, won a Nobel Prize in physics. He is now researching the evolution of language at SFI. Eric Smith, another accomplished physicist, is studying the chemistry that is most likely responsible for the earliest beginnings of life.* Brian Arthur, a highly respected economist, resigned his prestigious position at Stanford to do research into the evolution of technology. The policy at SFI, if you can even call it that, is to embrace everything that is interesting. The novelist Cormac McCarthy is a permanent fixture there, though I'm not sure what he's doing.

The reason that SFI is relevant to the people on Cox's little preserve

*In the second half of this book, I'll explain how Eric and other scientists now believe life came about.

is that when those visitors to the dunes made their mistakes—and when I made mine—we were all suffering from a disorder brought on by not being broad enough in our interests, by not being curious enough about our world and our position as people or animals in it. When we arrive here on earth, where we will spend all of the time allotted to us, we are naked, helpless, and ignorant. We are in a savage state, so to speak. As children we become brilliant generalists, curious about everything. But most of us gradually become specialists in our narrow little preserves, focused only on the minutiae of our own lives. We stop learning and then, when something unexpected happens, we don't know what to do. I think we can do better.

By having a broad range of knowledge, we can more easily change the frame through which we view our world and deal better with the difficulties we will inevitably face here. It is possible to go to the Outer Banks by way of an airplane and a rental car and to arrive at a rented house, as I did, without ever really knowing where you are. I had a route in mind but no sense of place. From the house, I went up the road to the state park and wandered around in the dunes. Without a general sense of the layout, I was already lost. I hadn't bothered to find out that the piece of land I was on was a long and narrow sandbar just off the Atlantic coast. You could go north and south, but there was water to the east and west. All I had to do to remedy this was to have the desire to learn. In some sense it doesn't matter what you learn, because you never really know what's going to come in handy. Geography class in school may have seemed boring, because we had no idea what good it might do us to know that subject. The answer is that we would know that the water to the west of the Outer Banks is not the Pacific Ocean, because the Pacific Ocean is three thousand miles away. (Yes, I already knew that, but someone didn't, according to Ranger Cox.) In part, our predicament is that we evolved to be well adapted to an environment that doesn't exist anymore, at least in technical cultures. We seem to live permanently in what Ranger Cox calls a "vacation state of mind," where all the old rules are suspended. In fact, most of us never knew those rules to begin with. We gradually evolved a culture that allowed us, as a people, to drop our guard. With the illusion that we have dominion over the earth, we conclude that we have nothing to fear.

The SFI style of addressing this state of affairs is to find everything interesting. Curiosity, awareness, attention—those are the tools we need if we hope to avoid our worst mistakes—and indeed if our children are to have a future on this planet. We have come to a pass in our evolution where we all must, to one degree or another, be scientists at heart or be victims of forces that we don't understand. I am certainly concerned about our survival as individuals. But I am also concerned that if we don't know the rules of our world—both the human rules and the physical rules—we will be in danger collectively as well.

Trouble will come to find us all. It always does. What form it takes, we cannot know. Age, illness, accident, terror, shattered love or business, the destruction of our global habitat by our profligate way of life—there's plenty of trouble out there to go around. We cannot keep it all at bay. But we have found a new kind of trouble lately. It's of a kind we've never seen before, and it has come upon us very fast. Our predicament is something like that of the people in the dunes: a bit lost, puzzled that we could lose our way in this commodious place. I remember my own reaction, wandering back and forth for an hour, thinking, But wait . . . I'm supposed to be having fun. This is a state park. I thought everything was arranged especially for my comfort and enjoyment here.

It was not. What did I miss?

Looking back over the history of science, I wonder why visionaries from Carnot to Kelvin to Clausius to Boltzmann could not have foreseen our predicament. The mathematics and science of what we were doing was within reach. One of them might have calculated what would happen if we burned not just the small amounts of coal that were being used in the steam engines of that time, but burned all the coal in all the world and all the oil we could lay our hands on as well.

The wonderful way of life that people enjoy in places like the United States is the direct result, beginning in the early 1800s, of an accelerating use of coal, oil, and gas. Everything in our technical culture that we enjoy today has its origins in the burning of hydrocarbon fuels. But, as Ranger Cox might have said, the rules of physics still apply, even here in this wonderful vacation land that we've created for ourselves. And in burning those fuels we liberated both the gifts they can bring and the penalty for their use.

This is not a book about global warming. It is about the unintended consequences, both small and large, of the interplay between human behavior and natural law. The natural mechanisms that make us behave the way we do, when coupled with the laws of physics and chemistry, bring about the short- and long-term outcomes that we both enjoy and suffer through. Each joy, each advantage we take, comes with its own cost. As we'll see, that is a fundamental law of nature. As a result, any discussion of this interplay between people and their environment leads directly to the costs. Global warming is the most notable penalty that has been levied against our way of life in recent history. We don't yet know what the ultimate cost will be.

On the other hand, there has been plenty written about global warming as a technical subject. What's more important, in my view, is how we get ourselves into such predicaments—and how we might get out. The same human behavior is at work, the same laws of physics, whether we're getting lost in the woods or facing a global crisis. My aim is to take into account the forces that shape human behavior and the laws that are fundamental to our world and, by understanding them, to create a new view of who we are and what we're doing here on earth. I think that change can come about only through a new way of looking at things. This is a book of changes.

In the first chapters of this book, I outline some of the ways we trip ourselves up and the mechanisms that carry us along through those tragedies. (I use the term in its theatrical sense. We are our own worst enemies at times.) I then look back into our human origins for clues to our sometimes baffling behavior. In the later chapters, I examine what's known about the origins of life and the universe in order to see if there are some broad principles by which we and all the physical systems we engage with operate. Here we begin to see the great unifying patterns and forces that govern our lives and the universe itself. Knowing those patterns, we can begin to see them all around us, and together we can marvel at them and ask new questions about them. We can also see how they have quite naturally led to our behavior over time and perhaps begin to understand better how that behavior has led to the predicament we're in.

When I was doing the research for my previous book, *Deep Survival*, I was attempting to learn how people get themselves into trouble and how they get out again. If we believe in a world of natural laws, there must be some underlying logic to the misery we bring upon ourselves and our triumph over it or our failure in the face of it. As I continued my research after the publication of that book, I realized that my interest had become deeper, broader, and more far-reaching. What if the mischief we make for ourselves is destined to result not just in the death of one individual or a small group but in a catastrophe of global proportions? I grew up under the very real threat of nuclear war, so global catastrophe has never been beyond my imagination.

But I don't think my place is to advise people on the practical matters of dealing with such threats. I leave that to the specialists. But it seems self-evident that we'd be better off understanding the rules that dictate our future. It seems self-evident, too, that if we don't develop a more accurate understanding of our world, we may bequeath a fairly bleak future to our children. And if, in the end, there don't appear to be pat answers to all the practical questions we now face, then at least we will have begun the process of changing the way we look at things. It is in that ability to have a different point of view that our best hope lies.

One

The Untied Knot

A few years ago, I was flying my airplane up to the Door County peninsula in Wisconsin on a beautiful summer day. Even after decades of flying, I had not lost the exhilarating feeling of my little aerobatics plane lifting off from the tiny grass airstrip among the cornfields of northern Illinois. The green world tilted away as I slid out over Wonder Lake and cruised north along the shore of Lake Michigan. I felt my nerves grow out into the stick and rudder, the wings and tail. I was one with the fragile tube and fabric and wood of the plane, as I surveyed the bluffs and the farmland, which sloped in waves of green, dotted with white cows, all the way down to the freshwater sea.

As I slipped north past Sheboygan and Manitowoc and Two Rivers, I became vaguely aware that the sky in the distance looked different from the rest of the bright blue world up there. It was dark. Very dark, in fact. During my preflight call to an FAA briefer, there had been no mention of bad weather. I had taken off in warm sunshine. I had reviewed my plan for the flight and created a mental model of how it would unfold, a sort of script that I could easily follow without thinking about it. And this dark patch of sky simply didn't fit that script. So I ignored it. I didn't make a conscious decision to ignore it. There was no deliberation involved. I didn't even have to push aside the new information; it just faded into the general background clutter of things to ignore. In my "vacation state of mind," I had easily dropped my guard. My stable mental model of the world before me, and my reliable script for what I was doing in it, guided me, as they always had, freeing my mind to think of other things.

I happened to be monitoring the air traffic control frequency in the Green Bay area, and I heard another pilot in a Beechcraft Bonanza report that he was in severe rain and thunderstorms. Given that I was flying in perfect conditions, this was simply too jarring to ignore. But even now that the new information had my attention, my first conclusion was that I must be picking up a radio bounce from somewhere far away. My first reaction was mild curiosity about this trick of electromagnetism. So even though it got my attention, it still hadn't disrupted my mental model. I was on my way to a vacation rental home where my family waited, and my model of the world easily accommodated this new element. I'd tell my daughters, Elena and Amelia, about it: "Hey, guess what. I heard a guy talking all the way from New Jersey." (Or wherever the Bonanza happened to be.) So I called the air traffic controller and asked where the Bonanza was.

When he told me, I felt my heart cease, then begin again. The Bonanza was just a few miles from me and dead ahead. In an instant my view of the world changed utterly. I understood what people mean when they say that the scales fell from their eyes. Everything seemed brighter, the colors more saturated, as adrenaline flooded my body and prepared me for those activities that attend the boundary between life and death. All at once I comprehended that I was doing something remarkably stupid. I was flying my tiny blue airplane with the happy stars on its tail into a black wall of explosive energy that extended from the surface of the water all the way to the heavens. At a speed of well over 120 miles an hour, I was closing with a force that I had no hope of surviving.

Fortunately for me, the Midwest is littered with small airfields, and I quickly radioed my intentions to an anonymous person at one of them, rolled the plane hard, and dove toward the earth. As I touched down, I gaped at that monumental wall of black, stitched up and down with lightning, as it swallowed the far end of the airfield. A line boy helped me push my plane into a hangar as the sky opened up with hail the size of marbles. Flying through that might well have killed me. The curious thing about my mistake is how unremarkable it was.

After the publication of *Deep Survival*, I went around the country talking to various groups of people (from firefighters and the military to institutional investors and survivors of cancer) about what might be called "risk management" or "decision-making." Sometimes we called it survival. But what we were really discussing was the embarrassing fact that smart people do some really stupid things from time to time.

One of the reasons for this has to do with the way the brain processes new information. It creates what I call behavioral scripts to automate almost anything we do. Behavioral scripts are an extension of the concept of mental models. Mental models have been widely studied and written about by psychologists for decades. A mental model can be something as simple as the image on a sign indicating where handicapped people can park their cars. We instantly recognize it as a wheelchair, even though it looks very little like a real one.

Mental models make us more efficient at processing information. Once we've seen a book, we will generally recognize all other books, no matter what the shape or size or color. Once we've seen a dog, we'll recognize both a Chihuahua and a Great Dane, though they look very little alike. Imagine how tedious life would be if we had to figure out such things anew each time we encountered them.

The fact that the brain creates these simplified models for quick reference explains why many optical illusions work. This is a common one, known as a Rubin vase after the Danish psychologist Edgar Rubin:

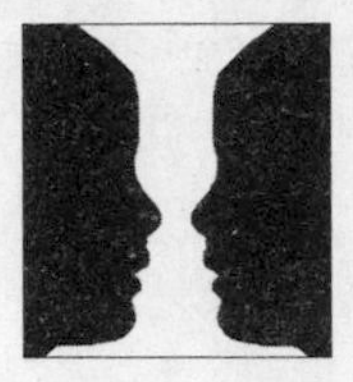

Photograph of Jonas Gonzales by the author

It can look like a chalice or like two silhouettes facing each other. It works because we have stored two mental models that are so much alike that our brains can't decide which is the correct one.

A behavioral script uses this same strategy of simplifying the world for efficiency (though using many different parts of the brain). For example, we can build a behavioral script for the physical motions involved in tying our shoes. Teaching a four-year-old to tie his shoes is quite difficult. But once he learns, he'll never forget. Practice causes a script to be written using new connections among neurons. Once written, that script makes the task so easy that it requires no thought at all. A behavioral script takes tasks that require all of our attention and transforms them into tasks that require none of it, freeing attention for other things. Behavioral scripts help us get around faster and with less effort.

A behavioral script will embody mental models, which identify objects in the world and the rules by which they behave. For example, if you want to put on your shoes, you don't have to examine every object in the room to see if it's a shoe. You will call up a mental model of a shoe, and your brain will automatically screen out everything you see until your eye alights on the shoe. Then your behavioral script will go to work to put the shoe on and tie it. You can be solving quadratic equations at the same time if you like. Your high-level thinking brain does not have to get involved with the business of the shoe. These clever scripts are run by the emotional system and deeper parts of the brain and body. Most of us learn a behavioral script that will cause us to duck (or catch) if something is thrown at us. A small child won't do this; the response has to be learned. An important feature of the script, however, is that when the right signal is present, the script will run on its own, without our consent, as it were.

Behavioral scripts are nesting structures in some cases. They can grow to incorporate other scripts and models to become very complex, as are the scripts for driving a car or skiing. Once you begin thinking about these models and scripts, you'll find them everywhere in your life. A behavioral script may be shared among various mental models. Once we know how to tie our shoes, we can also tie apron strings and bows on packages. A mental model may be shared among scripts as well. Once we recognize one brush, we'll recognize all of them. We may then use the model to create scripts for brushing our teeth, our hair, or our cat, all without thinking.

By a process of generalization or analogy, we may transfer a script from one type of action to another, and sometimes that transfer can lead to accidents. Once I was watching a painter work. He was drinking coffee while he worked, and frequently dipped his brush in a beaker of turpentine to clean it. Then, momentarily distracted, he dipped his brush into his coffee, and was just about to drink the turpentine when he recognized his mistake.

The fact that we all intuitively understand behavioral scripts makes them useful for sight gags. For example, in a Laurel and Hardy movie, we see Stan Laurel with a cup of tea in his left hand. Oliver Hardy asks him what time it is, and Laurel turns his wrist to see his watch and pours the tea on himself. How he could do such a thing requires no explanation. We all know we could do something equally silly.

Accidents of all types used to be analyzed in terms of their physical or mechanical causes. When the cause was clearly human error, they were often written off as the result of foolishness or lack of training. But among those who investigate accidents there is an increasing awareness that this type of analysis does not fully explain why otherwise rational people do what may seem irrational.

For example, in May 1989, Lynn Hill, the winner of thirty international rock climbing titles, was about to climb what she called a "relatively easy" route in Buoux, France. She threaded her rope through her harness; but then, instead of tying the knot, she stopped to put on her shoes. While she was tying them, she talked with another woman, then returned to climb the rock face. "The thought occurred to me that there was something I needed to do before climbing," she later recalled, but "I dismissed this thought." She climbed the wall, and when she leaned back to rappel to the ground, she fell 72 feet, her life narrowly saved by tree branches.

In her case, more training would not have helped. In fact, experience contributed to the accident. She had created a very efficient script for tying her rope to her harness. She could do it without thinking. But the script for tying her shoes was perilously similar to the one for tying her rope. The act of tying her shoes after threading the harness allowed her brain to reach the unconscious conclusion that her rope was tied, even while leaving a slight residue of doubt. It worked like the optical illusion of the Rubin vase. The system that did the learning was intelligent. But it led to the outward appearance of a stupid mistake.

In studying accidents, I've tried to go beyond conventional analysis to explain why seemingly stupid actions actually make a kind of biological and evolutionary sense at the time and under the circumstances. We can laugh at the Darwin Awards and write off our mistakes as stupidity or bad luck. But at some level, most of us are like Lynn Hill, with a knot half-tied somewhere in our lives, just waiting for us to put our weight on it. And one of the most frequently ignored factors in our behavior is the way we form models and scripts and use them rather than information from the world itself in most of what we do.

When I was a child in San Antonio, Texas, my grandmother had an ashtray that I loved. She had bought it in old Mexico. It was a ceramic rattlesnake, and its coiled body formed the tray for the ashes, while its raised head provided a menacing decoration. It was dark and dusty and very realistic. As I grew up, I always wondered what had become of that ashtray.

Decades later, I was hiking in the Santa Monica Mountains above Los Angeles. As I bushwhacked along a stream, I came across the ruin of a stone house. Everything was shattered and broken and overrun by thistles and weeds. Only the chimney remained standing. I thought I'd see if I could find a souvenir to take home, a bit of broken dish with a design on it or some old tool. As I poked through the wreckage, suddenly there it was: my grandmother's ashtray, complete and unbroken. Delighted, I reached out to take it. And then its tongue came out. I froze. Its tongue came out again. The serpent was smelling me. With the hair standing up on my neck, I carefully backed away.

This is an example of how our scripts and mental models can betray us. I had both a strong mental model and a familiar behavioral script working against me and blinding me to the obvious. Without ever considering it, I had been harboring a mistaken mental model of a rattlesnake. It was my grandmother's ashtray; it was ceramic and not real; it was not harmful. The nostalgic emotional attachment I had for it made it a strong and persistent model, difficult to displace. My lack of experience with real rattlesnakes made for a much weaker competing model. During that split second (about one-twentieth of a second) in which seeing becomes recognition, my brain had selected the strongest mental model, the ashtray.

The second process working against me was a behavioral script for what I was doing. I had told myself that I was looking for an interesting artifact to take home. That circumscribed and defined what I was doing. It formed the boundaries of my illusion, blinding me to what I knew intellectually to be true. If someone had magically snatched me out of there and questioned me, I could have articulated what I knew. I was in the dry mountain wilderness of California, where rattlesnakes are common. I was in a rock ruin, where rattlesnakes like to go because mice live there. Moreover, I could have stated one other obvious fact. The chance of finding an ashtray identical to my grandmother's anywhere in the world was close to zero. I didn't have any strong model or script (that is, direct experience) to support that intellectual knowledge. I had a strong script associated with the fun of finding something cool to take home from a hiking trip, because I'd done it many times. I had been rewarded in the past for following that script. Some of my favorite objects are found objects. Just as on my flight over Wisconsin, I was operating as if under a spell, in a vacation state of mind.

This kind of coupling of mental models and scripts leads to intelligent mistakes in all walks of life. When I reached out to the rattlesnake, I was exhibiting a flaw that attends many of our bad decisions: a fundamental lack of curiosity about the world around me. I was not paying deliberate attention. I wasn't intentionally pushing myself to gather new information. My behavioral script was sending a strong signal that said, "The world you see before you is a familiar one. Relax. You've got this wired." But as the acronym for behavioral script suggests, that message is often wrong. And this failure represented nothing more than the natural workings of the human brain. Over the eons, it has been good for survival to assume that what has happened before will happen again, and that what has not happened yet never will.

This assumption comes from the fact that the human brain classifies everything we encounter—along with the outcomes of our interactions with those things—and labels them good, bad, or indifferent. This system generalizes from previous experience. When things are similar, we lump them together into categories. We also tend to average things, ignoring special cases. We sketch out crude images for quick reference to tell us what we're perceiving and what its value is to us, if any. We create rules to go with those images, so that we know how things ought to behave, and we write scripts so that we can respond in a way that worked before.

When one of those scripts doesn't work and we are punished for our behavior, we revise our learning (if we survive) because we're smart. So when I at last comprehended that I was reaching for a real rattlesnake—in that moment of recognition, charged with adrenaline—I instantly created a new mental model for rattlesnakes and a new script for my behavior. The model of my grandmother's ashtray was gone, lost forever. And poking around in ruins would never be quite the same enjoyable pastime. Too bad. On the other hand, I'll probably survive longer, because I'd never try to pick up a rattlesnake again. The experience was a gift, because it awakened my curiosity about the world. I had an opportunity to cast off the mental models and behavioral scripts and explore what's really there.

When I speak of behavioral scripts, I am lumping together activities that brain scientists would classify differently. For example, tying your shoe would be called procedural learning, while searching through a ruin might have no name at all. But despite the fact that these activities involve very different parts of the brain, they represent an ancient and universal strategy: automating activities for the sake of efficiency.

Marc D. Hauser, an evolutionary psychologist at Harvard, wrote that "the cost of this kind of myopic learning system is that it will sometimes ignore the obvious." He trained a rat to navigate a maze in order to reach food. When he moved the food closer to the starting point, the rat ran right past it to get to where its model told it the food would be. When he put a wall between the rat and the food, the rat ran right into the wall without stopping. And when he cut off the end of the maze where the food was supposed to be and left a gaping hole, the rat ran right off into empty space. This illustrates the salient features of behavioral scripts. They allow us to ignore or discount new information (as I was doing on my flight over Wisconsin).

Daniel Schacter is a psychologist at Harvard with a special interest in memory. He had a patient he called Frederick who had developed a severe disorder that made him unable to form new memories. But Frederick seemed normal in many other ways. He could even play a decent game of golf. He knew the terms used in golf and even knew small details, such as the custom of placing a coin on the putting green to mark the position of a ball. But his amnesia was so severe that by the time Schacter had completed his putt, Frederick couldn't remember that he'd placed the coin on the grass. Nevertheless, before losing his memory, Frederick had played enough that the entire business of golf had become embodied in a very rich and complex behavioral script that had no need of Frederick himself. It could play golf quite efficiently—without him, so to speak.

These models and scripts form the basis not only of how we act but of what we perceive and believe. We tend not to notice things that are inconsistent with the models, and we tend not to try what the scripts tell us is bad or impossible. For example, it was long thought to be impossible to run a mile in less than four minutes. Doctors said it might prove fatal. Then, on May 6, 1954, a twenty-five-year-old runner named Roger Bannister ran a mile in 3 minutes 59.4 seconds. Within a month, another runner beat that record. By 1966, various runners had knocked 8.1 seconds off of Bannister's original time. By 1999, 16.27 seconds had been shaved away. Humans had not evolved into gazelles in those few years. What changed were the mental models and scripts of the runners.

Henry Plotkin, a psychobiologist at University College, London, calls this "the primary heuristic," a tendency to "generalize into the future what worked in the past." Past represents future. Whatever worked before, do it again. Whatever didn't work, avoid it. The brain is an organ of experience, from which it fashions generalizations and analogies. These form the underlying assumptions that shape our behavior. Without deliberately disrupting them, we are slaves to their dictates. Until Edmund Hillary climbed Mount Everest, everyone considered it impossible. Until Reinhold Messner climbed Mount

Everest without oxygen, everyone was held back by that belief, too. Today there are traffic jams on the summit, and climbers make it up without oxygen routinely.

Seymour Cray, famous for inventing the fastest super-computers in his day, liked to hire kids straight out of college, because they hadn't yet learned what was impossible, and so, unlike senior engineers, they went ahead and did it anyway. The same principle governed how Louis Leakey chose Jane Goodall, Dian Fossey, and Birute Galdikas to study apes. They had no formal training that would have convinced them that animals don't use tools, so Jane Goodall was able to make the breakthrough observation that chimpanzees do. Bill Atkinson is a legendary engineer from Apple who was largely responsible for the dynamic, easy-to-use Macintosh computer that revolutionized the industry. He first saw the "windows" concept at Xerox and was then able to quickly solve the problem of how to do it on his own. "Knowing it could be done," he said, "empowered me to invent a way." Mental models make our world, but they also shape and constrain the possible.

We can't see, or at least can't comprehend, things for which we have no mental models. The anthropologist Colin Turnbull once took a man named Kenge outside of the Ituri Forest in the former Belgian Congo, where Kenge was born and had lived all his life. The Ituri is so dense that Kenge had never had to create a model for the effect that distance has on the apparent size of things. The first time Kenge saw a herd of buffalo in the distance on the open plain, he refused to believe that they were not small insects. His brain did what brains always do and seized on the closest available mental model he had for something of that apparent size and shape. He could summon nothing else. In a real sense, seeing is not believing. On the contrary, believing dictates what we see. But once we can disrupt those models and truly see, then seeing becomes believing, and believing begets doing.

The efficiency that these scripts confer therefore comes at the expense of deliberate attention to real information coming to us from the environment. The model displaces the real world and sends this message: you already know about that, you may proceed. That's how many of our worst decisions are made. They aren't really decisions in the normal sense of the word. They're simply automated behaviors, formed out of the inheritance from our animal ancestry. And once a behavioral script is set up, we will continue to act until we're finished (for example, you can drive to the store and park your car while talking on the telephone) or until the illusion is somehow overturned by dramatic new information from the real world. For it really is an illusion, albeit one that is often useful. And these illusions on which we act can be very stable and difficult to overturn.

On November 17, 1999, a group of students at Texas A & M University were carrying out a tradition that went back to the 1920s. They were building a bonfire for homecoming. The final structure was made of felled trees that were placed on end and then bound with wire and stacked in tiers like a wedding cake. The pile was 40 feet high and weighed more than two million pounds.

What had begun in 1928 as a pile of scrap lumber had gradually evolved into a daunting engineering task. But no one had ever stepped back and asked: What are we really doing here? Is this a bonfire, or is it now something entirely different?

In the early morning hours of November 18, some seventy students were still working on the pile when it gave way, killing ten boys and two girls. Their mental models and behavioral scripts were very stable. They had been passed from generation to generation for the better part of a century. In many ways, the culture we now live in is like that. When ancient people lit their fires, they, too, were putting carbon dioxide into the atmosphere. But they didn't have the means to do it on such a large scale. Like the bonfire that kept growing, our ever-expanding use of carbon is now a towering structure, even as we continue to work on it.

My father, a professor at a medical school, always began his lectures with the words "Fellow students." He did so to remind his students—and himself—of the importance of continuing to learn. He would have liked the people at the Santa Fe Institute. He believed that his own field, biophysics, was too narrow. He professed an interest in everything. He believed that putting a large amount of diverse knowledge into his brain made it function better. As a young man, he would sit out on his mother's back porch in San Antonio, amid her oil paints and ceramics supplies, and entertain himself by solving complex mathematical equations. He tried to avoid the trap of looking at things as uninteresting just because they happened to be familiar. In time I realized that he was deliberately displacing his mental models, rewriting his scripts, in an effort to see the world anew. And to see the world anew is both to embrace it more fully and to guard against it. Confucius said, "A common man marvels at uncommon things; a wise man marvels at the commonplace."

Because of the way we make behavioral scripts, we are always rehearsing something, whether consciously or not. What we practice will be played back in some future setting. If the setting is different, the script may have outlived its usefulness. Most of the time, we aren't really aware of what we're practicing. The behavior that emerges may therefore surprise us. In his book *On Combat*, David Grossman describes a law enforcement officer who taught himself how to snatch a pistol out of an assailant's hand. During practice, he'd snatch the gun, then give it back and try it again. Finding himself facing a real assailant one day, he snatched the gun out of the man's hands, taking him completely by surprise. Then he handed it back. Luckily, his partner shot the assailant.

We are all, to one extent or another, slaves of our unconscious rehearsals. The rehearsals create the scripts on which we act without thinking. Our only recourse is to live mindfully, to learn broadly, and to be prepared to dislodge and revise those scripts. Doing that is not easy. But knowing how these human systems work can help.

To state the obvious, learning is a process by which we come to know something that we didn't know before. But what may not be as obvious is the fact that when we learn something new, that learning also changes the way we know everything else that we knew before. Learning not only changes the sum total of our knowledge, it changes the frame of all our knowledge. It alters our understanding of the world. And the person who learned is not the same as the person who set out to learn.

Two

Training Scars

Although the research into how the brain and body cooperate to shape behavior is still in its infancy, it appears that what I'm calling behavioral scripts are intimately involved with the limbic system—the emotional brain—along with other, more primitive, structures in the brain.* The chief characteristic of this system is that it pushes reason and logic out of the way and forces or facilitates action. We can see it at work in our everyday lives when we do something that we didn't intend to do, something that might have been accomplished better by thinking it through. My favorite coffee mug, when full, weighs a pound. Picking it up to drink has become a reliable behavioral script. I do it without thinking. If someone sneaked in while I wasn't paying attention and substituted a light Styrofoam cup in its place, I might run the script with the usual

*Discussions of the limbic system are becoming less common. It appears that brain and body work as one to produce rational thinking, emotion, and other functions that were previously thought to be located in one or another part of the brain.

amount of force and slosh the coffee all over myself. (Young children haven't formed the necessary behavioral script for this and so are given sippy cups as training wheels for learning Newtonian physics.) Only as I mopped up coffee would I have time to reflect, "How stupid of me." But it's not stupid. That sort of behavior, learned through practice and now automatic, is an essential feature of the way we've evolved. That system didn't have to work perfectly. It worked just well enough more times than not, and for millions of years it has been passed along. But now we have created a culture in which that same system can sometimes have disastrous effects. If spilling coffee seems too trivial, let's look at another example, because human nature doesn't use one strategy in one case and then invent a whole new one for another case. We carry all our automations with us wherever we go. You have to know about them and deliberately disrupt them, or they will simply play out whenever the right signals are present.

November 12, 2001, was a perfect day on the eastern seaboard with light winds and cloudless skies, as American Airlines flight 587, an Airbus A300, took off on a routine four-hour shuttle from Kennedy Airport to the Dominican Republic. The first officer for flight 587 that day was Sten Molin. The captain was Ed States. And as they sat at the gate, then taxied to the runway, their conversation reveals just how ordinary their circumstances seemed to them. They were discussing why seagulls congregate at construction sites. Captain States suggested that it was the coffee truck that sells sandwiches. Molin replied with a laugh, "It looks like a dump, that's why."

Molin, who would be flying that leg of the trip, was behaving strangely. He seemed to want to talk and started a conversation about the airline pilots' union. He was alternately yawning, laughing, and humming. At one point, he hiccupped and coughed. He was even singing as they waited to be pushed back. It left the impression that he was nervous about something.

At about five minutes to nine, the captain addressed the passengers over the public address system. "Well, ladies and gentlemen, Captain States again. We're all buttoned up and ready to go. We're just waiting for an airplane behind us to move on out of our way, and then we will be pushing back." Then the flight attendant began the briefing, which most of us have become conditioned to ignore. It all seemed completely routine.

As States and Molin waited, there was a ground crewman on the ramp with a headset plugged into the airplane's intercom to help coordinate the departure. At two minutes to nine, he told States that he was ready to unhook the voice link and start pushing the plane away from the gate. Captain States told him that the brakes had been released and said, "So long."

"So long," the ground crewman said, with no idea that he was saying good-bye to a man who would be dead in a few minutes.

In the meantime, States and Molin were going through the checklists, preparing for the flight, while carrying on their own conversation and conversations with ground control and others. Much of this activity would have become incorporated into behavioral scripts by then. At forty-two, Captain States had more than 8,000 hours of flight experience, almost 2,000 of it in military jets and transport planes. He drank socially, didn't smoke, went to the gym and to church, and was on the local Cub Scout committee. He'd gone to bed at ten o'clock the night before, rising at a quarter after four in the morning to go to work. He was a regular guy, well-liked and respected by his peers, and had faced no big emergencies in his commercial flying career.

Captain States was also comfortable with his thirty-four-year-old first officer. He'd flown with Molin on thirty-six previous occasions. Molin, nervous or not, was with a captain he liked and trusted. He knew he'd be back home the same day. He had every reason to regard this as a routine flight. Being younger, Molin naturally wasn't as experienced as States. He had logged only 4,403 hours of flying time, with 1,835 of it in the Airbus A300–600, the type of plane that he was flying that day. Before that, he had flown only smaller commuter planes. His, too, seemed a fairly ordinary life. The day before, he had helped a friend get her sailboat put away for the winter and had then gone out to lunch. That night, he'd had some people over for dinner and phoned his parents before setting his alarm for 5:30 and going to bed.

When they arrived to sit side by side in the cockpit that morning, Captain States didn't know that his copilot had somehow created a behavioral script that would cause him to do exactly the wrong thing in the face of a phenomenon known as wake turbulence. As the wings create lift, they spew air out behind them, leaving rotating winds in their wake. Those twin vortices slowly drift toward the ground and eventually dissipate. When made by a jumbo jet, they're forceful enough to upset a small plane or even damage it, but they will normally only jostle an airliner of the same size. I was once coming into LaGuardia in a 727 that hit the wake of the larger plane ahead of us. It felt as if it stood our plane on its ear, with one wing pointing at the ground and the other pointing at the sky—and it held us there for several seconds. From my point of view, it was quite dramatic. In reality, it probably tipped the plane about 30 degrees, and we were never in any real danger from it. Nevertheless, it will get your attention. It can frighten you, and when you're frightened, the emotional system takes over to initiate some form of automated behavior. A behavioral script will play out in whatever way you've rehearsed it. And it can make you do things that you shouldn't do.

It's impossible to know how Molin's script in response to wake turbulence came to be written in his nervous system. But the evidence is clear that it was there, ready to be activated by the right set of signals. It had been activated before, making it even stronger. In the past, two other pilots who flew with Molin had seen him overreact to wake turbulence in a manner that shocked them. In one instance, a captain for whom Molin was flying as copilot saw him get "very aggressive" on the rudder after encountering the wake of another plane. Molin "stroked the rudder pedals 1–2–3, about that fast," he testified to the

National Transportation Safety Board, which investigates all airplane crashes. Molin had pushed the rudder all the way to the stops, producing large and potentially damaging forces on the plane. That's exactly the maneuver that engineers at Airbus and American Airlines had worried that someone would attempt. It might spin the plane. It might break the plane.

It was the only time in his career that this particular captain had seen a pilot respond that way to a minor upset. The rudders are never supposed to be pushed to the stop during normal flight operations.* The captain in that 1997 incident warned Molin about it, but Molin did it again on a subsequent flight. The more you let a behavioral script play out, the more likely it is that it will play out again and the harder it becomes to stop it.

That same year, a flight engineer watched Molin respond to wake turbulence on approach to LaGuardia while in the clouds. After flying into the wake of the jet they were following, Molin slammed the throttles to full power, causing the plane to pitch up to an abnormally high angle. He then went around and shot the approach over again. There was no need to go around for another try at landing, known as a missed approach. A missed approach in the extremely crowded environment at LaGuardia could, itself, have put the flight at more risk. And the aggressive way that Molin handled the plane revealed just how stressed he was during what should have been an inconsequential event. What experiences, I wondered, could have caused him to develop such a deep fear?

At about ten minutes past nine on the morning of November 12, Molin was still yawning and chattering just before Captain States

*There was a dispute at the time about the use of the rudder in recovering from turbulence, but it was clear to the pilots that rudder was not to be used except during takeoff and landing at very low speeds. The NTSB cited no other reports of pilots doing what Molin did.

made this routine announcement: "Well, ladies and gentlemen, at long last, we are number two for takeoff toward the northwest today. Immediately after takeoff, we'll be in a left-hand turn heading for the shoreline and getting ourselves pointed southbound. About another two or three minutes it'll be our turn to go. Flight attendants, prepare for takeoff, please." A Japan Airlines 747 took off ahead of them at that point, leaving turbulence in its wake.

About two minutes later, the air traffic controller in the tower at JFK said, "American five-eighty-seven, Kennedy Tower, caution wake turbulence runway three-one left, taxi into position and hold." This was a standard warning that controllers give to everyone when a large plane has just taken off. Two minutes later, the tower cleared flight 587 for takeoff.

But Molin didn't take off. He looked out over the ocean, where the Japan Airlines 747, about whose wake the tower had just warned him, was angling away. And something stopped him from going. Instead, he turned to Captain States and said, "You happy with that distance?" Meaning: Is the 747 far enough away that we will avoid his wake?

It was an odd question. Molin was pilot-in-command at that moment. All decisions were his. As the pilot flying the plane, all he had to do was tell the tower he wanted to wait another minute to avoid the 747's wake—that's why the tower had warned him about it, so he could make his own decision. Instead, he asked the captain.

Molin's father, Stan, was a pilot for Eastern Airlines and taught his son to fly in a Cessna 152. He also sent him to a flight school in Tennessee to get his commercial license and the various ratings needed to embark on a career as an airline pilot. I couldn't help thinking about the son, a child flying with his father as teacher. I couldn't help wondering if he had ever asked that question: Dad, is this okay? Performing for the father is an extremely intense experience, no matter whether the father is authoritarian or benevolent or anything in between. Handling the father's stuff, his airplane, the symbol of all he is and does, is both a great gift and a burden for a child. Perhaps Molin couldn't think of himself as in command and so turned to his older and wiser captain.

Captain States seemed a bit surprised at the question, too, as he responded, "Ahh, he's . . ." and hesitated. Then he said, "We'll be all right once we get rollin'. He's supposed to be five miles by the time we're airborne, that's the idea."

But still Molin didn't go. He asked, "So you're happy?"

And again, I thought of Molin at the controls as a child with his father, the airline captain, watching him: Are you happy? I remember how it felt the first time my father entrusted me with the electron microscope in the lab where he did research. I was twelve. I was terrified of breaking it, and I wanted him to be happy with the way I used it. I remember, too, the first time I flew as pilot-in-command with my father, who had been an Air Force captain in World War Two. My father was a benevolent teacher, generous with praise, but it was still an intense experience.

It had now been a total of twenty-seven seconds since the tower had cleared Molin for takeoff, and while that might not seem like much, it's an extraordinarily long time at a busy New York airport. At last the plane began its takeoff roll. The JAL 747 was several miles ahead now, trailing twin cyclones behind it, one coming off each wing. Molin's plane was now accelerating toward them and would soon catch up with these wake vortices, as they drifted down and floated apart like the two sides of the wake behind a boat. If he had taken off as soon as he was cleared, Molin would have missed the wake of the 747 altogether. As it was, in his concern, he had delayed for exactly long enough so that he would fly across them at an angle.

But encountering the wake wasn't likely to harm him. Planes encounter each other's wakes all the time. If the planes are roughly the same size, it's not usually a serious matter. What was so crucial about this encounter was Molin's emotional reaction, over which he evidently had no control. His response was embodied in a behavioral script that didn't match the requirements of the real world. It was what police and military instructors refer to as a training scar, an inappropriate subroutine of behavior that somehow gets scripted during training, like the officer who grabbed the assailant's gun and then handed it back. Our lives are filled with bits of scripted behavior like that. Most of the time, they are of no consequence.

A little more than half a minute later, the landing gear were retracting, and they were on their way, with 251 passengers and a crew of seven flight attendants behind them. The fate of those people was now coupled to and controlled by the traces of flawed learning that had somehow been etched into Molin's system of nerve and muscle long ago.

During the next minute, Captain States retracted the flaps and slats and said, "Clean machine," indicating that they were set up for climbing to cruise altitude.

In response, Molin yawned and said, "Thank you." A yawn is an emotional response, not mediated by any conscious decision or controlled by free will. Stroke patients with a paralyzed arm will stretch with both arms when yawning, despite the fact that they can't move their arms voluntarily. Emotional responses come from a system in the brain that is separate from the one we use to control voluntary movements. Yawning can be caused by sleepiness or lack of oxygen. But it can also be a sign of fear, stress, or nervousness.

About fifteen seconds later, flight 587 encountered the first jostling of the 747's wake. Molin's response was a profound overreaction: he jerked the control wheel from side to side almost half its limit of travel. Big planes, like big ships, can't turn that fast, so there was no logical reason to move the controls that quickly or forcefully. No rational decision-making process could have led to that behavior. It was an automated emotional response, such as throwing up your hands to defend yourself when someone tries to hit you. It was the reaction of someone who was frightened.

In recent years, the NTSB has begun using a new technique to determine the level of stress a pilot is experiencing by measuring the fundamental frequency of his voice on the cockpit voice recorder. "Fundamental frequency" refers to how fast the vocal cords are opening and closing. The faster the rate, the higher the stress. The level of stress shows how much the pilot's performance is being degraded by fear. The analysts first measure the frequency while the pilots are sitting at the gate waiting to be pushed back. They use that as a baseline level and assume that it represents a resting or relaxed state. Then the instruments can track the change as the frequency gradually increases while the plane is being pushed back from the gate. As the plane taxis to the runway, stress increases with the increased workload. It should reach its highest level during takeoff, at which point a pilot who is functioning normally will be at what's called a stage one level of stress. That correlates with an increase of about 30 percent in the frequency of the voice and produces heightened attention and improved performance.

Beyond stage one, performance begins to deteriorate fairly quickly. If the frequency rises between 50 and 150 percent, stress moves to stage two, which results in "performance being hasty and abbreviated and thus degraded." Any further increase suggests panic, or stage three stress, at which point, for all practical purposes, you can't think or function.

The NTSB found that Captain States and First Officer Molin were within the normal range of stress, which increased during taxi and takeoff. But during that first encounter with wake turbulence, Molin's stress level shot up to stage two, while the captain remained calm. Captain States had no reason to be worried. The weather was good, the plane was good, and he had no fear of wake turbulence, even when they encountered it.

With Molin it was quite a different story. For whatever reason, he had developed a deep emotional response to wake turbulence and a behavioral script to go with it. And the initial encounter had put his limbic system on high alert, priming him for an even larger emotional response should anything else happen. It is a neurological fact of life that the chemicals of emotion fade away slowly. If you stub your toe, you're more likely to yell at your spouse, because your emotional system is primed; and the second response always tends to be disproportionately stronger than the first. This is why performers like to warm up an audience. It's why insults can lead to blows. It's emotional priming.

Then Molin flew into the second vortex—the other side of the 747's wake. His voice now showed so much stress that the normally dry NTSB transcript of it includes this note: "Spoken in strained voice." And what Molin called for in his moment of dread was "max power," though there was no conceivable need of it. It was the same thing he had done on the approach to LaGuardia in 1997, when he slammed the throttles open, causing the plane to pitch up.

By that time, Captain States was beginning to recognize that something was wrong, not with the plane but with Molin. The captain could see that Molin was jerking the control wheel back and forth with unnecessary roughness and must have thought it odd, prompting him to ask, "You all right?" Moreover, "max power" was completely inappropriate for the flight regime they were in. Once you're in flight, maximum power is used only in extreme emergencies, such as when an engine fails. Molin's was a panicked move and a panicked request. His was a flight response, as in "fight or flight." His behavioral script said, Get me out of here.

In a pressured voice, Molin answered, "Yeah, I'm fine." But he was not fine. Although Captain States probably couldn't see it because of controls that were in the way, Molin was forcing the rudder pedals to their stops, first left, then right, then back and forth again. And in doing that, he literally tore off the part of the rudder that moves. His frantic use of the controls had produced enough aerodynamic force to crack the fittings that held it on. Captain States didn't try to stop him, because he would have had no mental model to allow him to imagine that Molin could do such a thing.

With its steering gone, the plane veered over at a steep descending angle. Molin's response was a loud grunt. Then, as he wrestled with the now useless controls, he said, "Holy shit." The demon that had been stalking him for so long had hold of him at last. He'd made the thing he feared the most come true.

In the cabin, the passengers would have heard the loud bang as the rudder broke away. They would have seen the abnormally steep angle of the plane, the ground rushing up from their relatively low altitude of 3,300 feet. As the airplane accelerated, it reached the speed at which the pressure of the air itself will tear a plane apart. Passengers behind the wings would have seen the engines rip off and go tumbling away. Then clouds of white fuel sprayed out of the broken fuel lines and caught fire, engulfing the wings and fuselage in orange flame. Every airline passenger's worst nightmare was coming true.

In the cockpit, the cacophony of chimes and stall warnings was being drowned out by the roaring of the air stream, leaving a space of a few seconds in which Molin, still believing that this was all because of wake turbulence, exclaimed, "What the hell are we into? We're stuck in it."

Captain States, perhaps now convinced of that, too, urged him, "Get out of it. Get out of it."

Those on the ground saw a great silver vessel spinning out of the sky, towing a rooster-tail of flame. It was caught on tape by a video camera at a highway toll booth. It was a terrifying and oddly beautiful sight. One of the engines destroyed a boat that was parked in someone's driveway. And when the airplane itself hit the ground in Belle Harbor, New York, it not only killed those on board, it completely destroyed four homes and killed five more people, who most likely hadn't even considered the hazards of flight that day.

American Airlines flight 587 lasted just over a minute and a half.

The saddest thing about this story is that if Molin had done nothing, nothing at all, in response to the wake turbulence, the flight would have proceeded safely to the Dominican Republic. The final report from the NTSB put it this way: "In fact, if the first officer had stopped making these [rudder movements] at any time before the vertical stabilizer separation . . . the accident would have been avoided."

There is a common type of car crash in which a single car rolls over on the highway for no apparent reason, killing or injuring those inside. After years of investigation and with the help of a few survivors, it's become clear what happens in such crashes. For whatever reason (sleepiness, inattention), the driver allows the car to drift onto the shoulder. Suddenly realizing the mistake, he reacts just the way Molin did, with an automated response mediated by fear, a behavioral script. He jerks the wheel to get back into his lane. Most of us rehearse this response, unaware, in city traffic at much slower speeds. Someone pulls in front of us, and we jerk the wheel to avoid a collision. It works well most of the time. The emotional system unconsciously tags it as a good response to the world. Small scripts easily nest within larger ones, and that little script for behavior becomes folded into the larger one that we've developed for driving. Unfortunately, jerking the wheel at highway speeds can roll almost any car. New situation, old response.

In most of these cases, it's helpful to remember that when smart people do stupid things, it is the process of learning that is intelligent, not the action of the person. The learning is only as good as the environment in which it took place. Once things change, new learning is needed. What's smart depends entirely on context. When Molin made his mistake in the Airbus, there was nothing wrong with the airplane, just as there is nothing wrong with the cars that roll. In the parlance of the NTSB, the accident resulted from an "aircraft–pilot coupling event." More specifically, it resulted from coupling a maladapted behavioral script to a large aircraft containing 259 other humans.

In the early days of aviation, the spin was a mysterious event, a death spiral from which pilots rarely recovered. Knowing that, a pilot who found himself in a spin would bail out if he happened to be blessed with a parachute. And then people began to notice something strange. After the pilot bailed out, the plane would sometimes right itself and fly on until it crashed or ran out of fuel. A clever pilot proposed this: the airplane wasn't at fault. The pilot was doing something that kept the plane in the spin. Remove the pilot, and you solve the problem. Pilots began learning how to recover from spins by doing less, not more.

Recognizing such a coupling event is a step in the right direction for understanding and perhaps avoiding some of the accidents—and other situations—we get ourselves into in life. We are complex systems ourselves, armed with a host of these behavioral scripts and perhaps unaware of many of them. And when we interact with other complex systems, human or natural or machine, those training scars can become greatly exaggerated. That's why knowing ourselves and our world can be so important.

Three

When Modern People Face Ancient Hazards

As a species we've become tremendously successful. And that very success has become incorporated into the generalizations and analogies created by our emotional systems. Our models and scripts are based on some underlying assumptions that may lead us astray at times. For example, most people who are likely to be reading this book will always have enough to eat and drink. We always have adequate shelter. We rarely even think about predators. We haven't been forced to be very curious about our world in order to discover a strategy for survival. We haven't had to pay close attention in order to gain any little advantage from our environment. All advantage flows to us effortlessly.

The central task for the emotional system is to classify the consequences of our behavior in the world, so the natural conclusion the organism draws is that we're doing something right. If we're getting fed every day, our strategy must be a good one, so we don't even have to think about it. We can simply continue doing what we've always done. Life is good. Our little corner of the world is safe. But life wasn't always like that. It still isn't for many people in the world. The emotional systems that those people build are very different from ours.

After the tsunami in the Indian Ocean on December 26, 2004, Indian officials discovered that all the 250 or so members of the Jarawa tribe on the Andaman Islands had survived. Those people live in the Balughat forest, and, through DNA samples, they have been traced back 60,000 years to their evolutionary roots in Africa. Their folklore and legends contain warnings about where to go and what to do when earthquakes occur and especially when the sea retreats from the beach, as it does when a tsunami is on the way. They went to high ground and stayed there for several weeks, surviving on coconuts. These wise people run all but naked in the forest, and their highest form of technology is the bow and arrow. At least 900 much more technically advanced people were killed by the tsunami on those same islands. (Unfortunately, the Jarawa and related tribes—some of the last of the original people on earth—won't survive the encroachment of newer people like us.)

Man was once such a wary creature, venturing onto the plain with his head down and his ears up. The world must have seemed calculated to bedevil him, to snatch from his grasp all that he held dear, and then to distribute his bones to the beasts just for spite. It must have seemed the height of good sense, then, to craft a world where the fruit always hung low and all predators were kept at bay. Because we are so very clever and have big brains and highly versatile hands, we have done just that. Short of an airplane crashing into my house, there is nothing in my immediate environment that will reach out to kill me. But as a result of that great achievement, we have evolved a vacation state of mind, a culture that teaches us to drop our guard. While the Jarawa tribesmen were using good sense and the learning they'd accumulated about their world over many generations to survive the tsunami, hundreds of thousands of people from much more technical civilizations died in the face of it. Because of our clever tools, we have a rich history of those events in videos, and that history shows us the sad story of how helpless we can be at times.

One video, shot by Anukul Charoenkul, owner of the Viewpoint Restaurant at the vacation resort of Khao Lak in Thailand, shows a lone man facing the sea as the wave approaches. He doesn't move. He just stares at it until it sweeps him away. Other videos show crowds of people on the beach, watching the wave but doing nothing to escape. One of the most haunting videos shows people on the beach at Panang, milling about, walking casually, or standing in relaxed attitudes, hands on hips, as the great wave approaches in the background. They seem completely unaware. One of them is obviously recording the video. They all see the wave but do not move away. When the wave reaches them at last, they laugh as it gathers around their feet and ankles and begins to rise. Only when the wave knocks them off their feet and starts to sweep them away do they scream, as they comprehend the grave miscalculation they've made.

Not everyone had an opportunity to survive. Chance always plays a part in catastrophe. But many who could have escaped did not, because nothing in their experience had prepared them for an event that, in geologic time, happens routinely where the land meets the sea. Educated and sophisticated as those people may have been, their mental models and behavioral scripts were useless when their environment underwent a completely predictable change. They had created a stable mental model of their world and an indelible script for what they were doing. They were on vacation in the benign sunshine of a happy beach. Only at the last, as they were knocked over, did the wave sweep away that model and rewrite that script, in many cases too late to do them any good in the future.

I corresponded with one man who survived. He was from California and was vacationing on the beach in Thailand with some friends. The friends had stayed up drinking all Christmas night, while he went to bed. The next morning, he was up early and felt the earthquake happen. It was a faint rumbling, far off. (The earthquake happened about 500 miles away, and the waves took about an hour to reach Thailand.) But he was familiar with the earthquakes in California. That experience had made him wary. He, too, had a behavioral script for being on vacation. But it was much less stable. The faint but recognizable rumbling of an earthquake upset it and put him on his guard. As he ate breakfast on the beach in front of his hotel, he was alert, curious about what was changing in his environment. He was casting about for new information. He recognized that his way of conceptualizing his environment and what he was doing—eating breakfast, on vacation—might need to be revised. It was as if he had dropped back a million or so years in evolutionary time to engage his Paleolithic mind, which I will examine in the next chapter.

Then he saw the sea retreat, sucking and hissing, leaving live fish flopping around on the sand, until the water had vanished for a distance of three-fourths of a mile. That was enough to forge a new model and to script a whole new behavioral strategy. The new model included the obvious way that waves work: what goes out must come back. He reasoned that it would come back with a vengeance and a deliberate intent to take his life. The new script was to flee. He rushed inside, woke his hungover friends, and led them to the fourth floor, where they remained on a balcony as the waves deluged the city, killing thousands.

If the Indian Ocean seems too remote to be relevant to us, consider another area that is vulnerable to tsunamis: the Pacific Northwest. Earthquakes happen where two slabs of the earth's crust collide. The Juan de Fuca plate is one such slab. It is being driven beneath the North American plate across an area that stretches from northern California up into British Columbia. The Makah tribe in the area of the Straits of Juan de Fuca have a legend to remind them what to do in case of a tsunami. It tells of a time when the sea withdrew, leaving the beaches dry, and then returned to submerge the entire country except for the mountaintops. The legend also explains that some of the nearby islands became populated when people took to their canoes during a tsunami and were carried there by the receding waves. The legend contains warnings to head for higher ground when the earth shakes or the water recedes. For the people living in the Pacific Northwest, the warning time will be short, because the subduction zone, where the earthquake will occur, is right off the coast. A tsunami took place there in 1700 and was carefully documented by the Japanese. The wave spanned the entire Pacific Ocean and hit Japan with waves measuring 16 feet. By analyzing the Japanese data, modern Canadian researchers have been able to determine that the earthquake of January 1700 measured 9.0 on the Richter scale and ruptured the boundary of the two plates from northern California to Vancouver Island. The waves would have been considerably higher in the Pacific Northwest, perhaps even larger than the highest waves from the tsunami of December 2004.

A government warning from the Provincial Emergency Program of British Columbia describes the situation today: "A similar offshore event will happen sometime in the future and is a considerable hazard to those who live in southwestern B.C."* *Will* happen, not may happen. The native people of Vancouver Island have an oral history of the earthquake of 1700, in which they describe the tsunami, landslides, and destroyed villages. Their stories also note that it was winter, and it took place before any Europeans had arrived.

You don't have to be a genius or a native running naked in the woods to let learning help you. On the morning of the tsunami in Thailand, Tilly Smith, ten years old, who had come with her family from England to vacation on the beaches there, saved the lives of about one hundred people. At ten years of age, your models and scripts are constantly being revised and so are not as stable. Tilly had been introduced to tsunamis in geography class and chose to perceive and believe. She was paying attention. Because of her alertness and the flexibility of her models, she connected new information from her environment with what she had learned and turned it into a new

*Although it's common in the popular press to say that we're overdue for an earthquake when one hasn't happened in a long time, that's not how such events work. They're clustered in time, not periodic. That means that when you've waited a long time for an earthquake, you'll probably have to wait even longer.

model of her world. "I was on the beach and the water started to go funny," she told reporters. "There were bubbles and the tide went out all of a sudden. I recognized what was happening and had a feeling there was going to be a tsunami. I told Mummy." The fact that her parents listened to her, believed her, and took action based on what she said suggests that they, too, are remarkable people, willing to change mental models on a moment's notice.

Events like these are not as remote and irrelevant as we might like to think. They can reach right into any of our lives. The tsunami reached my own quiet suburban neighborhood. One of my neighbors, a successful real estate salesman in his thirties named Ben Ables, happened to be on vacation in Thailand and was found among the dead.

In the Shinto religion there is a story told of an emperor who was riding his horse through a thunderstorm. He passed a house where a cat was sitting on the porch, and the cat waved to him. The emperor thought this was so unusual that he went over to see what had made the cat wave. As he dismounted and walked toward the porch, lightning struck his horse and killed it. Cats became revered in Japan for their protective powers after that. The way of worship in many religions is a ritual form of paying attention.

When Mount St. Helens began erupting in May 1980, the local county sheriff, Bill Closner, tried to warn people to leave the area. He even tried erecting barriers to keep them out. But no matter what he did, sightseers came in droves. Entrepreneurs were selling maps that showed how to avoid roadblocks and use logging roads to get up the mountain. Closner and other law enforcement officers tried to keep the owners of cabins in the wilderness out of the area, too; but the people appealed to the governor of Washington, who agreed under pressure to allow them in. As Sheriff Closner said later, "As it turned out, the worst thing that could have happened did."

Like the beaches of Thailand, Mount St. Helens had been a popular recreation area. The previous year, half a million people had visited the Spirit Lake area alone. Nothing bad had ever happened to them there. The volcano became active at the beginning of the tourist season. People had made plans. They had packed their bags and spent money. They had fashioned models of the previously benign place to which they were going and scripts for what they were going to do there. Those models would be hard to disrupt. Any sort of investment in an intended goal—of work or money, say—makes it harder to change direction.

By May, the tourist season was in full swing, and, like the tourists on the beaches of Thailand, many people simply couldn't conceive of the events that were about to occur. It wasn't in their repertoire of experiences. The intellectual knowledge of what Mount St. Helens was going to do was readily available. Experts were quoted on the news, talking about it every day. But not even that could displace the sense of conviction that our models and scripts can sometimes confer.

Embodied in those models and scripts is the lesson that we take from modern life, the culmination of many eons of hope and hard work by people across the globe. It is the ultimate human dream come true, to fashion a world that is truly our own, bent to our wishes, there to serve and delight and protect us. And since we implicitly trust our analogies because of our success, when a new situation arises we find it easy to fashion an analogy that seems to fit. We're on vacation just when the mountain is set to blow. This is going to be a great show. And perhaps this, too: Didn't we see something like that at Universal Studios . . . ? We're used to calamity being presented to us in a safe context, on television or in the movies. Just as I associated a rattlesnake with the warm feeling of being at my grandmother's house, we train ourselves to associate catastrophic events with the enjoyable experience of being entertained. Our natural wariness has been disabled by false experience.

The evidence of their peril was all around them: smoke, fire, and the trembling earth. The people ignored all of that and not only stood around to watch but ventured nearer despite police roadblocks. Some were even camped out on the bridges across rivers that ran through the valley. They brought coolers of beer and laid out picnics. They couldn't conceive of something that could kill at a distance of many miles. They had missed out on a critical type of learning. It was not their fault. They were not stupid. They were simply far out of their element, just as a Jarawa tribesman would be in a modern city. The difference is that the Jarawa tribesman would know that he was no longer in the Balughat forest. The vacationers at Mount St. Helens believed they were still in a world that they controlled. The concept of being out of our element is difficult to come by in our culture, because we've fashioned an artificial environment in which all elements seem to be ours.

At 8:32 a.m. on May 18, an earthquake measuring 5.1 on the Richter scale peeled away the north flank of the mountain, releasing the largest landslide ever recorded during historical times, more than half a cubic mile of material. The landslide carried away all the trees and much of the rock, splitting along three pathways near the bottom, devastating the Spirit Lake area, and running up a 1,200-foot-high ridge five miles to the north. It could not be outrun. The largest mass of material moved at 150 miles an hour, covering the Toutle River valley floor with debris to a depth of 500 feet in places. In all, some 24 square miles were buried.

When the side of the mountain fell away, it released the tremendous pressure that had been building, causing the volcano to blow rock, ash, and hot gases sideways to the north, destroying everything within 150 square miles. No tree was left standing for 6 miles. All the lesser vegetation was blown away within a further 13 miles. The explosion was heard 700 miles away.

At the same time, the volcano fired a column of ash many miles into the sky. The effect of the ash moving so rapidly through the air produced enough static electricity to touch off lightning storms, which in turn started hundreds of forest fires. The ash was dense enough to produce a false night for 125 miles across the plains as it moved east. (It nearly killed a planeload of people when it clogged and shut down all the engines of an airliner.) Then some 540 million tons of ash began raining down over 22,000 square miles. An eyewitness in Spokane told me that it piled up in the streets like gray snow.

In the early stages of the event, lava erupted at speeds of up to 470 miles an hour. As the upward spray of molten rock fell back into the crater of the volcano, it spilled out and over the edge, flowing down the mountain. This so-called pyroclastic flow, moving at 145 miles an hour, picked up snow, ice, and debris to form lahars, dense flows of sediment and water with the consistency of wet concrete. More than 65 million cubic yards of this material were deposited in one drainage alone. Scientists measured the temperature of the pyroclastic flows at more than 1,200 degrees* (660 C), about the melting point of aluminum. Of the fifty-seven people killed, twenty-three were never found. The only reason more tourists weren't killed was that it was still early on a Sunday morning.

Even then, some people couldn't believe. A film crew sneaked past the roadblocks a few days after the eruption, were severely burned, and had to be rescued by helicopter.

Past experiences that reward our behavior (or simply fail to punish it) make our scripts and models feel reliable. For example, a long road trip in the mountains equals big fun and great photos to take home. Certain objects, people, or sensory inputs can set familiar scripts in motion—say, a scenic overlook that signals to us that it's time to stop and take photos.

*I will use Fahrenheit throughout, followed by Celsius in parentheses, unless otherwise noted.

These scripts, like computer scripts, can have modules in them that require us to stop and think. For example, once we get out of the car, we'll have to set up the shot and manipulate the settings on the camera. This linear way of thinking, which otherwise might help us, is being carried out in the service of the script and, as a consequence, makes the script more stable rather than disrupting it. We place our attention on an element required by the script, not outside of it to check whether there is important new information coming from the environment, or whether this might be the wrong script. We tighten our view rather than broaden it, leading us to miss important cues.

Ready at last to complete the script and take the photo, we tell our spouse, "Okay, just a little farther back now . . . Smile."

Dozens of people have fallen to their deaths off the rim of the Grand Canyon at the scenic viewpoints. According to park records that date back to the 1920s, more than 20 percent of them died while taking or posing for photographs. Their model of what they were doing could not have been farther from reality.

These models always contain within them unconscious or unstated assumptions that may be wrong. Those types of assumptions are often the reason that certain kinds of puzzles are hard to solve. Here is the classic "nine dot" puzzle. The instructions say to connect the dots with the smallest number of straight lines without lifting your pencil from the page.

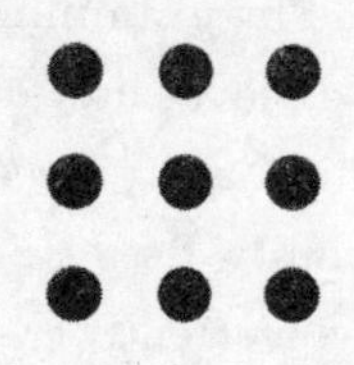

Most people will see that figure as a box made by the three dots on each side surrounding the dot in the middle. They start with the assumption that they have to work within the box. (This puzzle has been suggested as the origin of the phrase "thinking outside the box.") But that illusion will lead to a solution that requires five lines. Here is a solution with four lines:

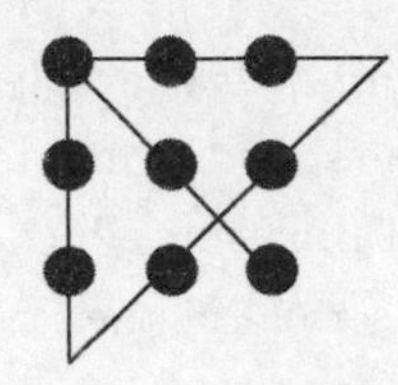

The box is not a box at all. Like the Rubin vase, it's an illusion resulting from the way the brain makes mental models. A box is simply the most familiar mental model suggested by the arrangement of dots in the first illustration. And once the solution is drawn, it's difficult to see the box any longer, just as it's difficult to think about photography when someone has just fallen into the Grand Canyon.

In many ways, our behavior as a species on this earth has been a long process of following such illusions through our lives and passing them on from generation to generation in a slow and incremental process that has led us to this point in history, where we face the toughest question of all: Are we capable of changing our traditional, inherited ways of behaving before they lead not just to personal catastrophe but to a world that is no longer so hospitable to our kind? If we are to change, we have to understand this question more deeply. One way to do that is to see where we came from and what sorts of problems we were attempting to solve for ourselves long ago. Where we came from tells us the evolutionary trajectory that shaped who we are and how we behave. And by examining how we solved our most pressing problems, we can discover how our overall strategy for survival led us to this point, which, as we'll see in subsequent chapters, is a turning point for all humanity. Because in solving the problems of our ancient ancestors, we inadvertently created new problems that no species has ever faced before. As a species, we are now facing sticker shock over the price of our own salvation.

Four

The Sign of the Hand

To my mind, one of the most remarkable discoveries in history is a set of footprints made by our ancestor *Australopithecus afarensis* at a place called Laetoli in Tanzania. Discovered in 1976 by paleoanthropologists working for Mary Leakey, the footprints show that two, possibly three, people were walking together across a field of freshly fallen volcanic ash about 3.6 million years ago.

Not far away, a volcano now known as Sadiman had erupted, and a cloud of fine ash had settled on the landscape. Rain began to fall, perhaps because the eruption shot ash into the air and produced static electricity. The ash turned to a paste, like plaster, and as the people crossed the field, they left their footprints.

The footprints show one larger individual, a smaller one, and possibly a third who may have walked in the footprints of the first. The smaller set of prints was made in such a way as to suggest that the person might have been carrying something, perhaps a mother with a child on her hip. Others have suggested that she was pregnant. Because of what we know about how humans and the other apes* behave, I like to imagine that this was a small family: father, mother carrying a child, and possibly another child walking.

Although it's not possible to know such things with any certainty, I imagine a mother holding her baby as they fled the eruption, beset on all sides by beasts much larger and far better equipped to deal out death than she and her family were to defend against it. Throughout the area, impressions were made in the same fresh ash by elephant, rhinoceros, hyena, scimitar cats, false saber-toothed tiger, giraffe, antelope, giant porcupine, woolly mammoth, three-toed horse, and short-necked giraffe, just to name a few. I wonder what the Laetoli woman would have been thinking, her heart beating fast, the flush of adrenaline, the quickening steps. While I wouldn't want to attribute too much brainpower to her, I've seen chimpanzees and bonobos behave in such intelligent ways that I feel confident that the Laetoli woman was very clever, very aware, and able to make at least short-term plans and execute them.

The Laetoli woman can help us learn about ourselves. She can help us to think differently about who we are and why we behave the way we do. Learning about our ancestors (and our nearest relatives, the other apes) will change the frame through which we view the world and how we envision our place in it, so that the next time we are watching the ocean rise up to smite us or witnessing the earth smoking and trembling, we will think of the Laetoli woman. Who are we and how did we come to behave as we do? What deep ancestral currents still flow through these modern veins of such ancient design? How much of our love, hate, violence, kindness, and our clever fabrications belong to the Laetoli family, too? Pondering—if not answering—these questions can help us. And I firmly believe that we need the help.

When I think about us now, watching a great wave advance or

*Humans are a species of ape in the family *Hominidae*, which includes orangutans, gorillas, chimpanzees, and bonobos.

standing flat-footed before an exploding volcano, my mind goes back to the Laetoli woman, and I wonder how she coped with her world and what sort of journey brought us from there to here. On that day, as ash rained down from a darkened sky, I see her hurrying on her way, clutching the most precious thing in her world: her child. Her child embodied the fate of a species, the DNA, which alone represented the possibility of a future. Only by safeguarding that precious package could she and her kind go forward. Even if she could not conceive of that, she had built-in systems and had learned automated behaviors that would have encouraged her to get her child to safety. That system would have made her more likely to succeed. Even though she could not imagine this place where I now live, she was unconsciously striving toward it even then. It was as if she could grope through the darkness toward such a place but never quite fully comprehend it. It was a place I'll call Safetyland, the world we've fashioned to serve us. There was no such place then. For the Laetoli woman, there would be only the urgent drive toward a promised land that she would never reach. Hers would be a temporary refuge, and she would be carried there by a script unconsciously drawn from experience for her own survival. Yet it is in her ancient urges that we find the earliest origins of the world we've crafted for ourselves.

Where did the Laetoli woman go after the eruption? What did she do? Perhaps she was headed for a cave where she'd found shelter before. Perhaps she envisioned the protective cover of the forest that grew in Tanzania at the time. Maybe she carried something else, too, a stone or stick, against the beasts that might threaten her. The behavioral script that led her on was an emotional map of sorts, learned from her elders and her own experience. It was a chemical road, paved on one end by adrenaline and cortisone and neurotransmitters, which gave her a bad feeling brought on by fire and smoke and loud noises and beasts. That feeling would motivate behavior by way of a chemical called dopamine, among others. It would lead her gradually to a place full of endorphins and oxytocin, where cortisone and adrenaline would fade away, leaving only the good feeling of warmth and safety in the comfort of her family at the end of the journey. That's all they would need on their biochemical journey: to move from bad feeling to good feeling by carrying out the actions that had worked in the past.

That, in essence, is how we still act and make decisions and move through necessary tasks. We have exactly the same chemical road that the Laetoli woman had. Of course, we have much richer thoughts, and we have language, too. We also have the ability to think by way of step-by-step reasoning beyond our automated scripts. But most of what we do isn't moderated by deliberate logic. It takes shape out of the same natural systems that drove the Laetoli family across the open field.

Wherever the family went, Sadiman erupted again after they had passed, and another layer of ash fell, covering their footprints. It rained again and the ash turned to paste and dried, hiding and protecting the footprints until a few million years of erosion revealed them at last.

Unfortunately, we can't really know much about the Laetoli woman. We don't have her bones. But we do have Lucy's bones. Like the Laetoli family, Lucy was an australopithecine. She lived in Ethiopia about 3.2 million years ago. She was not genetically descended from chimpanzees or bonobos. All apes share a common ancestor, from whom we split off about seven million years ago. But it may be useful to imagine Lucy's intelligence and abilities as lying somewhere between those of chimps and our own, though much closer to the chimp than to us.

It appears that the absolute size of the brain may not matter as much as the ratio between the size of the brain and the body. The size of Lucy's brain and body had about the same ratio as a chimp's. She probably weighed 60 pounds and was in her late twenties or early thirties when she died. She would have had superior abilities to develop complex mental maps of the area where she lived in order to cope with its hazards and take advantage of its resources. She would undoubtedly have used some form of simple tools, as chimps and bonobos do, and she would have had the brainpower to maintain complex social relationships. So Lucy and the Laetoli family were very smart. The fact that people of their kind survived among the wild beasts and erupting volcanoes for a million years across the more than 1,000 miles of terrain that separated them tells us how very smart they must have been.

We are lucky enough to have about 40 percent of Lucy's skeleton, including most of her right hand and some of her wrist bones. Mary Marzke, an anthropologist at Arizona State University, has spent her life studying hands. She is an expert on figuring out what hands can and can't do. She and her colleagues developed a 3-D computer program that can analyze the surface and volume of bones in ways that were impossible just a few years ago.

Marzke points out that Lucy could grip with what is known as a "three-jaw chuck"—the grip a baseball pitcher uses. Lucy also had the ability to form a good pinch grip, such as you'd use to turn a key in a lock. She might not have been able to make a five-jawed cradle grip, such as you'd use in holding this book with one hand, thumb on top. If you make a claw of your hand and hold it as if you're going to scratch your own eyes out, you will see how the fingers radiate out at angles from the center of your palm. The other apes can't do that, and Lucy probably couldn't either. Their thumb and little finger don't rotate the way ours do, which gives us the ability to make a five-finger chuck and a good power grip. That allows us to use one stone to hammer accurately on another. In that ability lies the very earliest beginning of the highly technical culture we enjoy.

Among many other attributes, we share Lucy's pelvis. By studying the places where muscles attached to the bone, Marzke has learned that Lucy could throw overhand like a human. Pitching that way relies not just on the hand but on the entire arm and shoulder, as well as on the ability to rotate the hips in coordination with both shoulders. The reason that a baseball pitcher can throw the ball at 100 miles an hour lies in the intricately coordinated movements of the hand, arm, shoulder, spine, and hip. Lucy would not have been as good as we are at this kind of throwing. No other animal, including Lucy, can throw as accurately as we can. She didn't have the curving spine we have or the hip joint that would allow her kneecap to face completely forward, as ours does. These two features give us our uniquely fluid style of walking. Although she walked upright, Lucy probably would not have been that graceful. But she would have represented a great advance over chimplike creatures.

Because of their special brains and hands, Lucy and her Laetoli cousin may have been capable of making rudimentary stone tools, but none has ever been found. All we know is that somehow, in that hostile place, they managed to survive without the fangs, claws, size, or speed of other animals. All they had was intelligence and very clever hands. I believe that Lucy and the Laetoli family must have made tools, but that they were probably made of perishable materials such as wood. There is a group of savannah-woodland chimpanzees at a place called Fongoli in Senegal. They sharpen wooden sticks with their teeth and use them as spears for hunting. They cool off in pools of water, inhabit caves part of the time, and even keep pets.

Our opposable thumb allows us to make all the grips needed for both precision and power. Ulnar opposition gave us the ability to hold a stick straight out in alignment with the arm. No other ape can do that. That ability, in turn, gave us a prodigious advantage in being able to club or stab something (or someone). As Frank R. Wilson, a neurologist at the University of California and Stanford schools of medicine, put it in his book *The Hand*, "Having the ability to telescope the arm outward would convey a lethal advantage." (At the age of about fourteen months, human children begin pointing; chimps do not.)

Lucy, in other words, can be seen as poised on the edge of becoming human. The demands of throwing and the new uses to which her hands could be put in wielding rocks or other tools would have challenged her 400-cubic-centimeter* brain to struggle with increasingly difficult problems, such as tracking a moving target if she wanted to hit it with a stone—say, for hunting. The emergence of humans as apes with very large brains was intimately tied up with the increasingly complex uses to which we put our hands.

There were other pressures on the brain to grow as well, and it is in this growth of brain and hand that we find the origins of our present culture. Not all parts of the brain grew at the same rate. The frontal cortex was especially favored by the explosive growth that the brain would undergo in the next million or so years. This is the part of the brain that we use for logical, stepwise thinking, the part that we so intimately associate with being human. The neocortex, which includes the frontal lobes, is a double layer of brain cells that covers the rest of the brain like a rind. It would be about the size of the Sunday funnies if you unfolded it and laid it out flat. The front part of the neocortex will cause us to question the moronic things we do, from wrecking the car to running off with the nanny. And while it's very useful in many ways, it may not always be able to stop us from doing such things, even when we recognize them as stupid.

Another demand placed on that part of the brain involves society. The larger an animal's social group, the larger his frontal lobes must be in order to keep track of individuals. And humans form very large groups of up to 150 people.† This is related to reproductive strategy,

*Five cubic centimeters is about one teaspoon; 1,000 ccs is about a quart.

†Robin Dunbar, an anthropologist at the University of Liverpool, has shown that the ratio between the size of the neocortex and the size of the brain determines the number of individuals an animal can know well enough to form a social group.

among other things. There are many strategies for reproduction, such as spewing out half a million eggs, as oysters do, and hoping for the best (if oysters can be said to hope). But people, like the other apes, have very few children. An oyster can lose lots of offspring. We cannot afford to. So if our strategy is going to work, we have to take care of the children we have. Taking good care of the kids means being smart, which in part means having a bigger brain.

Being smarter also means having time to learn how to use that large brain, which means having a long childhood. Donald Johanson, the anthropologist who discovered Lucy's bones in Ethiopia, described this process: "The best way to learn during childhood is to play. That means playmates, which, in turn, means a group social system that provides them. But if one is to function in such a group, one must learn acceptable social behavior. One can learn that properly only if one is intelligent."

So there were many ways in which our brains were influenced to grow larger. Moreover, all of these influences were mutually dependent. One didn't cause the next in a linear progression. They all worked together in a catalytic fashion. After a million and a half years or so of development along these lines, then, the brain of Lucy's descendants had at least doubled in size in the person of our most recent and direct ancestor, one or another of the species we call *Homo*. There are many names assigned to individuals who might represent our most recent ancestors. Among the names you'll see are *Homo erectus*, *Homo heidelbergensis*, and *Homo ergaster*. As my friend Kathleen Ratteree, a graduate student in anthropology at the University of Wisconsin at Madison, told me, "The world of paleoanthropology is highly contentious. No evidence is certain and no lineages are clear."* For convenience, I'll refer to our most recent ancestor as *Homo erectus*.

Whatever you choose to call them, we do know that Lucy's descen-

*For a detailed discussion, see *The Last Human* by G. J. Sawyer and Viktor Deak.

dants became more and more like modern humans over time in both the size of their brains and the abilities of their hands. While primates going back to the earliest creatures in trees were able to manipulate objects with their paws, true hand control, such as Lucy possessed, marks a qualitative change in both the brain and the body. One of the pioneers in this field of study, Merlin Donald, said, "Hand control involves, for the first time in evolution, a coming together of visual, tactile, and proprioceptive feedback* on the same action system. Hand control may be regarded as the crossing of a biological Rubicon." Sometime between *erectus* and about 200,000 years ago, when the earliest of our kind of humans appear, a completely modern hand developed, and the size of our brain had expanded to its present-day average of 1,300 cubic centimeters.

As the anthropologist Ralph Holloway at Columbia University pointed out, the need for sequential planning and mental rehearsal that throwing requires encouraged the growth of the brain and especially, once again, the frontal cortex. Throwing accurately requires extremely fast and intense processing of information, which in turn requires a huge amount of computational power, along with an ability to imagine the future in great detail. For instance, in throwing a rock or a spear to kill an animal or a person, we must first conceive of and connect these objects, actions, and events: hand throws rock; rock kills animal; man eats animal (or person). Then we must calculate with great accuracy such quantities as the distance to the target; the tension in hundreds of muscles, ligaments, and tendons; the weight of the stone; the distance by which we must lead the target if it's moving; and the amount to elevate the trajectory to compensate for gravity. Even with a computer, this is not a trivial task. Yet we are very adept at throwing, and at least some of us may be born to it.

Frank Wilson put it this way: "The shoulder, already the heart of

*This is a fancy term for the mechanism that lets you touch your nose with your eyes closed. Through proprioceptive feedback we know where our limbs are in space.

an overarm propulsion system for moving the body through the trees, was admirably suited to transfer its capabilities to the as yet uninvented world of ballistics." The advent of accurate throwing and the ability to make and use spears would bring a bonus of protein to feed the rapidly growing brain. The climate was changing at the time, making resources more scarce. This would have increased the need to keep track of those resources through the creation of more complex mental maps of their geographical location. That, too, would have put new demands on the brain.

The hands themselves embody a great deal of who we are, both individually and as a species, and they carry tremendous symbolic significance. (Some of the earliest art consists of hands painted or outlined on rock walls.) When the company once known as Enron fell in disgrace, Ken Lay, Jeff Skilling, and other of its executives were arrested and handcuffed with their hands behind their backs. The police didn't think they were violent criminals in need of restraint. They were top corporate executives. No, we as a society bound their hands as the ultimate symbol of their loss of status as humans. You have lost your hands. You have been cast out of the group and are now a lower animal.*

When we wave to someone, we show our palm, signaling that we are of the same species, a species that can wave, and showing that we are unarmed. Similarly, when we shake hands in the American way, grabbing each other's hand firmly, it is a symbolic message of mutual disarmament. You immobilize my hand, and I immobilize yours. By this mutual consent, we have agreed, for the moment, not to kill each other. In a different culture, like that of my Indian ancestors in Mexico, whom we'll meet down the road, a light touch of the palms is enough to speak the sign of greeting.

---

*Obviously, this concept would not apply to someone who had lost his hands in an accident.

Now I think back to the Laetoli woman, crossing that field of fresh volcanic ash millions of years ago in Tanzania. The ash would have felt like warm beach sand. The air was choking with smoke and soot. Lightning illuminated the landscape briefly, and then the gloom closed in again, as deep thunder rolled around the shores of the lake where she lived. The land was teeming with predators. She was a small woman, less than four feet tall. Her mate would not have been much larger, her child that much smaller. They walked barefoot and wore no clothes. All they had were their minds to carry them through.

When Sadiman erupted almost four million years ago, it may well have been the first time the Laetoli woman had ever seen a mountain explode in fire and brimstone and the sky grow dark with a rain of ashes. Her existing mental models may not have given her much guidance in shaping her response. But she would have had at least some rudimentary ability to use new information to make plans and then execute them in a sequential fashion. So at that moment of truth, as the Laetoli woman and her mate crossed the grim landscape, she would have been using whatever limited aptitude she had for sequential processing to try to update her existing behavioral scripts, all in the service of moving her as efficiently as possible to safety. And that little extra bit of thinking of which her species was capable resulted in its surviving that hostile environment.

That ability to think is also related to the development of her hand and arm and the complex capacities it would eventually make possible, such as making tools and throwing accurately. The neurological machinery for throwing was laid down in such a way that the skill could be learned easily and early and then transferred to simultaneous processing—throwing without thinking. The frontal lobes must continue to play a role in selecting a target and analyzing conditions. Significantly, the frontal cortex must also restrain throwing until that analysis is done and the right moment arrives. Then the higher brain must let go and transfer control to the emotional and lower parts of the brain so that the throw can proceed flawlessly. We can see the beauty of that when a baseball pitcher is doing well. We can see it break down when action takes place without the frontal cortex (panic) or when the frontal lobes won't let go (clutching).

In Western culture, we tend to believe more in stepwise logic. But if you rely solely on that, then you miss out on a rich source of information from your environment. Our modern brains are our great advantage over Lucy. But completely rejecting that older kind of thinking can sometimes hurt us.

In the summer of 1994, a forest fire was started by lightning just outside of Grand Junction, Colorado, in a mountainous area known as Hell's Gate Ridge just above the Colorado River where it flows along Interstate Highway 70. Because of a drought, there were many fires to fight, and this one was neglected for some days before firefighters were put in place on top of the ridge above it. It had seemed like a small, inconsequential fire, but by the time the firefighters arrived, it had spread and was growing fast.

The team was led by a smoke jumper named Don Mackey, who had more experience as a firefighter than as a leader and decision-maker. While he couldn't put his finger on any logical trail of evidence to support his position, he felt that the team should leave the ridge. He had an intuitive sense that conditions weren't good and that being above the fire could put them in danger. More than once he voiced that opinion but didn't take decisive action to leave. Mackey could sense that something was wrong. That is what's known as simultaneous processing at work. But the information he was receiving was not in a form that could be processed by the high-level, conscious part of his brain that might have allowed him to make the decision through the stepwise logic of the left brain. The angle of the terrain, the direction of the wind, the color of the smoke, the heat or light or the absence of birds and wildlife—so many things combined to create his impression of the hazard that it was impossible to synthesize it in any logical way. In bureaucracies, such as the U.S. government, the one that Mackey worked for, simultaneous processing is usually ignored, while so-called sequential processing, with its clear explanations, is required. So Mackey followed his established models and scripts and kept going in, deeper and deeper, even while he was worrying out loud that the fire was not worth getting hurt over. Bureaucracies force us to practice nonsense. And if you rehearse nonsense, you may one day find yourself the victim of it.

At the same time that Mackey was going deeper into the fire, another firefighter, Byran Scholz, arrived and "felt a pinprick of apprehension," as John Maclean wrote in *Fire on the Mountain*, a harrowing account of what happened. Scholz had seen the very same type of situation a few weeks earlier and had watched the fire overrun an area of half a mile in the space of a minute. Scholz couldn't logically explain what he knew, either. The cues and signs he picked up were too subtle and gave rise not to logical explanations and clean lines of evidence but to gut feelings instead. Like the visitor in Thailand who woke his friends after feeling the earthquake and seeing the tide go too far out, Scholz's previous experience had made him willing to revise his models and scripts, and he now acted decisively based on the information he was receiving through simultaneous processing. Mackey radioed him, asking him to send some of his crew down to help, and Scholz refused, moving his crew to safety instead. When a 150-foot-high wall of flame came roaring up the ridge, Don Mackey and thirteen others died. Scholz and his crew survived. (He could not have saved Mackey and his crew by going down into the fire with them. Joining Mackey would have done nothing but increase the number of casualties.)

I've worked with firefighters over the years, and I heard of one captain who makes a habit of setting the alarm on his watch to go off once an hour whenever he's working a fire. When it goes off, he stops to look around and question what he's doing, what he's missing, what he ought to notice. Moreover, he stops to consider what he's feeling in his gut, too. Maybe there's a signal he's ignoring. Maybe he's thirsty and hasn't let himself notice. If he is, he stops his crew and makes them drink water, because dehydration is often the first step in setting us up. It impairs judgment and can allow other factors to organize themselves into an accident. But most important, that captain is unconsciously casting his mind back to Lucy and using the whole range of abilities of his marvelous human brain.

The discussion of how our brains grew so large won't be complete without considering language. Most of us will have noticed that when people, especially children, are doing something that requires precise and difficult use of the hands, they also tend to make movements of the tongue and mouth. (Michael Jordan did this when he performed his famous "Air Jordan" flights across the basketball court.) Various researchers have demonstrated that some of the same general areas of the brain control both the sequential organization of grammar and the sequential steps needed to make and use tools. The mouth and hands share some of the same neural landscape as well. A number of researchers have been moved to wonder if sign language came before spoken language or in conjunction with early vocalizations. The demands of a precision sign language and of making tools may then have put pressure on the brain to grow, at the same time that precise movements of the tongue and lips, the mouth and throat and diaphragm, began to give rise to more complex spoken language. The hand, tongue, and brain together may have formed a synergistic circle of reinforcement, as each spurred the others on to do more and grow more complex and competent. Frank Wilson put it this way: "Evolution has created in the human brain an organ powerfully predisposed to generate rules that treat nouns as if they were stones and verbs as if they were levers or pulleys. . . . [We] build sentences the way we build huts and villages."

Steven Pinker, the director of the Center for Cognitive Neuroscience at MIT, among others, credits the hands with the creation not only of language but of the underlying mental abilities as well. As he put it in his book *How the Mind Works*, "Hands are levers of influence on the world that made intelligence worth having. Precision hands and precision intelligence coevolved in the human lineage, and the fossil records shows that hands led the way."

Because of the brain's natural ability to change and adapt, this process would not have had to take very long. As Jeffrey Schwartz, a research professor of psychiatry at the UCLA School of Medicine, wrote, "the basic anatomy of the adult brain, not just the details of its wiring, can be altered by the demands its owner places on it." Since this was first demonstrated in 2000 by researchers studying cabdrivers in London (the map-making hippocampus grows larger in the brains of cabdrivers), numerous other researchers have confirmed the fact that profound changes in neural functions can begin within a few days. For example, the sense of touch will take over the visual cortex in people who learn Braille after going blind. This rapid adaptation could help to explain why some evolutionary developments, such as art, appear so suddenly without any apparent antecedents. And once an innovation appears, it may also spread very rapidly, because all apes imitate one another. We tend not to try what we consider impossible. Once some genius among us has demonstrated that it is possible, we find new abilities in ourselves. Think of Everest and the four-minute mile. Language may have marked one of those sharp, evolutionary transitions.

On the other hand, there is new evidence that the ancestral seeds of sign language may have existed in apes for millions of years. Broca's area on the left side of the brain has long been associated with the production of speech in humans. Part of this area, known as Brodmann's area 44, was discovered in chimpanzees and gorillas. The researchers observed that apes use right-handed gestures along with vocalization as a means of communicating. The sign of the hand and spoken language seem to have very deep roots. Indeed, they are so intimately bound up that it's possible to observe a child's use of gesture at age one and extrapolate from that to his ability to use language at age three.

If humans began communicating with more and more precise and complex gestures, they would have been elaborating the system of nonverbal communication that involves facial expression, body language, scent, and other subtle signals. Whether through long evolution or through an abrupt conversion, this would have led to much finer motor control in the use of those gestures and in further changes in the hand. Several forces, then—such as sign language, tool-making, vocal utterances, and an ability to plan structured and complex stepwise behavior—could have been working synergistically. A grammatical language would have fit in nicely with, and would have worked in support of, those abilities.

In the case of sign language, one hand (subject) performs the actions (verbs) while the other serves as the object of that action. If you point your left index finger at the sky and grab it with your right hand, you can see how you're making, in effect, a sentence with the structure subject–verb–object: "Hand grabs finger." As Pinker wrote, "it is a significant discovery that . . . languages have verbs, objects, and pre- or post-positions to start with, as opposed to having the countless other conceivable kinds of apparatus that could power a communication system." And despite the number of ways languages could order the elements, 95 percent follow the ordering system we use or its inverted form, as in Japanese.

Once the tongue started to be used for language, it required the brain to make a decision, in effect, on which side of the brain would control it. Otherwise, a kind of tug of war would ensue (which is what happens in some people who stutter). Both the dominant hand and the power of speech settled in the left hemisphere. The tinkering instinct, our love of manipulating objects, helped refine the sequential thinking associated with the left brain, which was well-adapted to making a grammatical language. The long chains of thoughts we are able to have are a reflection of the long chains of physical movements we learned to do long ago in the systematic manufacture and use of tools, as well as, perhaps, in sign language. We still use our hands for communicating, and gestures can sometimes be more effective than words. Physical exploration of the world underlies all of what we think of as cognition and intelligence. As Oliver Sacks, a clinical neurologist at Albert Einstein College of Medicine, put it, "This process . . . cannot arise, cannot even start, unless there is movement—it is movement that makes possible all perceptual categorization."

Small children incessantly talk to themselves about the events of the day in a stream of consciousness. They're consolidating memories and even honing skills. This sort of activity helps prepare us to predict the future and our own reactions to new events. It is part of our process of making models and scripts.

The brain treats words as concrete objects, because the concrete objects of the world exist only as mental models in our brains. The same part of the brain handles word and hand. Words that can conjure and manipulate those models are as real as the objects themselves. Through this process, then, our words become our world. Our innate ability to create and call up and alter mental models is what creates metaphor. Chimpanzees who have been taught sign language can make metaphorical leaps. I wouldn't be surprised if *erectus* could, too. *Erectus* didn't leave us any record other than his bones and tools, but because of them we know that he was the first human being to travel far and wide. We do have records from more recent early human culture, since the advent of oral and written traditions, when we made long journeys on foot. Language provided a distraction during those onerous trips, even while serving the useful function of synchronizing the emotional state of the group. The issue of whether or not *erectus* could talk is contentious, but perhaps some sort of language helped to carry him through his long journeys as well. Since his kind were the first of us to spread widely across the globe, I am forced to wonder if and how language and sequential thought might be related to those migrations.

When we read about the origins of humankind, the story is often told as if everyone evolved in a sort of linear lockstep. If that view were true, then everyone you know would be a chess master or a Mozart. Or a moron. Certainly there was as much variation in talent and skill among the kin of *erectus* as there is today among modern humans. Once our cleverness had earned us the luxury of idle time, those differences in talent and skill would have become that much more pronounced. So it must have been with language. Most people would have functioned on an average level in terms of how well they could communicate. A few words and signs would suffice, and the average skill was good enough for the day-to-day workings of the world. But a few could use the language as a transformational power in the tribe—orators, poets, entertainers. Like chess masters, who memorize the patterns of pieces on the board, the poets memorized thousands of familiar patterns that allowed them to improvise long stories in order to spontaneously incorporate new events and render them meaningful to the tribe.

People still do this. One modern scholar recorded an improvised poem sung in the oral tradition that still exists in what used to be Yugoslavia. It was 12,000 lines long (this book is about 8,000 lines long). We can now imagine a very early time, perhaps more than a million years ago, when a process like this may have been carried out through dance, gestures, perhaps even words, to achieve emotional transcendence and meaning for the group. When language soars, it allows us to have as our own the experience of another. As with ancient art, it can put our eyes into another's skull and our heart into another's chest.

I have to wonder how *Homo erectus* would have accomplished certain tasks without language. Along with the complex problems of coordinating a hunt, the evidence suggests a pressing need for high-level communication. F. Clark Howell, the legendary anthropologist from the University of Chicago, discovered evidence of these hunts in Spain in the 1960s. Donald Johanson, his student, describes it: "It is all there to be seen: the stone tools the hunters left lying around, the traces of burned grass, the animal fossils." These early humans formed hunting parties that would carefully light the grasslands on fire in just the right way so as to drive herds of elephants into swamps. As the animals wallowed in the mud, trying to escape, the hunters could move in and kill them. Just setting a fire to direct the elephants rather than scatter them would require an exceptional ability to communicate. The complexities of the hunt at that point would have been daunting enough with language and would seem impossible without it, even if it was mostly body language, facial expression, and hand signals.

Clark found one elephant that had been butchered. The people had carved up half of it and taken it away, leaving the other half in the swamp. This evidence shows remarkable restraint, which suggests a significantly enlarged frontal cortex. Other apes eat whatever they find or kill on the spot. Restraint in the presence of food represents a significant change in behavior. The restraint that the frontal cortex exerts is one of its essential features. Indeed, a hallmark of being human is restraint. Eight of the Ten Commandments concern what not to do.

The scene of the hunt has far-reaching significance and says a great deal about what sort of person *erectus* was more than a million years ago. The intelligence and imagination that it would have taken to come up with such a hunting strategy sets this creature far apart from other animals. The strategic use of the combination of fire and mud is a dazzling stroke of genius. The ability to then attack and kill an animal the size of an elephant and to create, on the spot, the tools with which to butcher it—further, the ability to coordinate the crew to carry the meat home—were highly sophisticated undertakings that mark the beginnings of our present-day prowess and our present-day predicament. For without such intelligent use of technology, there would not have been a ready supply of food. Without the food, the human species could not have proliferated in such great numbers. If the human species had not proliferated in such great numbers, we could not be doing the damage we're now doing.

Somewhere between Lucy and *erectus*, we found the wherewithal to make better decisions inside the parentheses of restraint. Gut impulse, instinct, intuition, simultaneous processing, as we've seen, can be very important elements of decision-making. Decisions can't be made in the absence of emotion. But to operate mindlessly is to live at the mercy of chance. We need both reason and emotion, both sequential and simultaneous processing, to be fully aware. We needed some form of reason and directed attention to get from Lucy to where we are now. We'll need them even more to reach our own future, if we're to have one.

On that long journey from Laetoli to here, did we take a wrong turn somewhere? Surely many of us have learned much and become a great deal smarter in many ways. Why then does it seem that our intellectual powers grow weakest just when we need them most? How did we come to drop our guard like those blithe vacationers at Mount St. Helens? It happened by increments, small steps in our efforts to find food more effectively and ensure our own safety more reliably.

Five

Groupness

Whatever their level of intelligence, Lucy and the Laetoli family knew one thing with absolute certainty: trouble would find them. And when it came, it would be big trouble of the kind that can end your life and that of your children and stop the species in its tracks. The trouble they faced was not the kind where they sat everybody down to talk it over and take a vote. This was the kind of trouble where they had one chance to act, right now, and the choice had better be the right one. They did not have to know this in any cognitive way that would allow them to express it. They knew it globally, internally, deeply. There was no vacation state of mind for them. They never dropped their guard.

But we have to leave Lucy and the Laetoli family behind now. Their lives—what little we can know and imagine of them—are instructive, but they were too far from us in time and there isn't enough evidence. Nevertheless, we gain a better understanding of our own behavior in the highest reaches of our modern technical culture if we recognize how very close to our ancient relatives we still are. Decisions in big business, world politics, and even the high-tech world of space flight are predicated on traits and innate biological and social systems that we developed millions of years ago. Whether we can transcend that ancestry is an open question. We certainly have no hope of doing so if we don't know anything about it.

Because of that, one of the things I've done in my research is to spend as much time as I can with the largest colony of bonobos in captivity, which happens to be in the Milwaukee County Zoo, about an hour north of where I live. For a number of years, I've been going there to spend time with them. Although I had read about chimpanzees and bonobos, there's nothing like seeing them firsthand to get a feel for their intelligence and behavior. True, they are in a zoo and not in their natural habitat, but so are we. Barbara Bell, who is in charge of their well-being, invited me to come and watch from the wooded area behind the public enclosure at the zoo. There the bonobos can go outside to be away from the crowds of people who come to see them.

Chimpanzees and bonobos are quite different from each other. Both chimpanzees and bonobos fight and have sex, but chimpanzees fight more, and several males will mate with any female who comes into heat. They don't have sex when the females aren't in heat. Bonobos have a reputation as the "hippie chimps" that make love not war, but that's not quite accurate. Bonobos are capable of extremely violent behavior. They sometimes literally tear each other new assholes, an expression that probably came from the way apes tend to fight. Nevertheless, bonobos are more likely to settle their disagreements by having sex. That tends to reduce the instances of violence. As most marriage counselors will recognize, when couples don't have sex, they fight.

I've spent long hours in the forest in front of the bonobo enclosure with nothing but a fragile-looking chain-link fence between me and them. Sometimes, as I watch them, a 325-pound gorilla named Hodari paces back and forth behind me, beating his chest to get my attention. (Hodari likes people.) He, too, is kept in by nothing more substantial than a chain-link fence, and I don't doubt that he could break through it if the thought ever occurred to him. I've watched bonobo couples sit and French kiss for hours on end, seemingly just to enjoy necking with each other. I've seen them have sex in every conceivable combination. They use sex in all sorts of social situations, such as relieving stress or forming an alliance. No female needs to be in heat for them to copulate.

Chimpanzee society is dominated by males. Bonobo society is dominated far more by females. The most dominant individual in the Milwaukee clan is completely bald over her entire body, because everyone wants to groom her. They've simply picked her clean.

Chimps are more likely to be violent than bonobos. In March 2005, two chimpanzees in a private zoo called Animal Haven Ranch, about 30 miles east of Bakersfield, California, attacked a man named St. James Davis and nearly killed him. Davis and his wife, LaDonna, had raised a chimp named Moe in their home. Chimps tend to become more aggressive as they age, and Moe had bitten a woman and was put in the zoo in 1999. The Davises had brought a cake for Moe's thirty-ninth birthday. Two other chimps somehow escaped and attacked them. For all the time they'd spent with Moe, the Davises didn't realize that by accepting food, Moe would be violating a primary rule of ape society. In addition, their mental model of him as a pet prevented a deeper analysis of what they were doing.

In ape societies, including human ones, status is everything. Status dictates your access to resources, such as food and the ability to reproduce. Status is life. But what's most important is not necessarily having high status. It's knowing what your status is and behaving accordingly. As a newer member of the group at Animal Haven Ranch, Moe did not have much status. He therefore had no right to the cake, nor to the attention, which should have been given to a chimp with more social status. Roger and Deborah Fouts are researchers who have worked with chimpanzees for decades. They run the Chimp and Human Communication Institute at Central Washington University. Deborah Fouts told reporters after the attack, "Chimpanzees have a real sense of right and wrong and fairness and unfairness."

The two escaped chimps, Buddy and Ollie, attacked as Mrs. Davis was cutting the cake. One of them bit off her thumb. Her husband pushed her behind a table to protect her and then took the brunt of the attack. Chimpanzees and bonobos, as Barbara Bell told me, "like to go for the digits and the butt." They also go for the genitals and the face. Buddy and Ollie bit off Davis's nose, cheek, and lips. They bit off all his fingers and chunks of his buttocks. They castrated him and gnawed off part of his foot, breaking his heel bone. Davis survived only because someone shot the chimps.

There are frequent fights among the bonobos at the Milwaukee County Zoo. Barbara has tremendous influence with them. She provides all their food and manages the population, ensuring that vulnerable individuals are separated from others who might harm them. Nevertheless, with limited space and complex social relationships among the twenty-five or so bonobos in her care, she can't prevent every fight. She recalled coming in one morning and finding one of the younger females with a finger nearly bitten off and hanging by a flap of skin. Many times she arrives at work to find a bonobo bloodied.

When I first started observing the bonobos, there was a new member of the clan, Viaje, who had been rescued from Mexico, where he'd been kept isolated in a cage. He was at a potentially fatal disadvantage. He was not a member of the group and therefore was fair game for anyone to kill. In addition, he had not been raised in bonobo society, so he didn't know the rules. It is no favor to an ape to teach him human ways unless he's actually human. Fortunately for Viaje, Barbara's careful intervention kept him from being killed, and eventually he learned a few rules and began to be accepted.

It was not easy to work Viaje into the group. "But we had to have him," Barbara told me. "He was caught in the wild, and we really needed to breed him." Populations of apes in zoos can go stale very quickly in terms of their genetic diversity. New DNA is especially hard to find when it comes to bonobos, because they're so rare and can no longer be imported from their native habitat in Congo, where they're being rapidly pushed to extinction by people who hunt them for food and by the destruction of their habitat. A bonobo from the wild is therefore a real prize. The trouble is, being the new member of the group, Viaje had no right to mate. Taking his pick of the females would have been a killing offense. So it took some wily machinations for Barbara to get Viaje alone with the right females without getting him killed.

The females, as well as the males, will try to kill any male who tries to breed if he doesn't have the appropriate status. Barbara put Viaje together with a "founder female," as the animals are called when they've been caught in the wild. Maringa was only thirty four years old, but she'd been crippled by a neurological disorder, so she wasn't as dangerous as the others. Viaje successfully bred with her, and a baby named Faith was born in February 2005. "Now," Barbara told me, "we need another girlfriend for Viaje, somebody who will be nice and not rip him to shreds."

Even after Viaje had been fairly well integrated into the group, he suffered from his low status. I watched one day as he sat on a hammock hanging high in the air while an immature female, daughter of one of the leading females of the clan, repeatedly flew around the cage, rapping him painfully on the head each time she passed. Viaje had to sit there and take it. The older female had her two sons in the enclosure with her, and if Viaje had laid a hand on the young girl who was tormenting him, her mother would have given a subtle signal—almost imperceptible to humans—and her two big sons would have torn Viaje to pieces. Viaje now understood this, having been badly beaten more than a few times. Whenever Barbara gave out bananas as a treat, Viaje hung back from the fence, unwilling to take one. It just wasn't worth the beating. (Barbara would slip him one later.)

When thinking about how the emotional system and our mental models and scripts operate to help or hurt us, it's useful to think of ourselves as two creatures instead of one. One of the creatures is the everyday person that we rely on to make decisions, hold a job, balance a checkbook, or follow a recipe in a cookbook. The other is either a chimpanzee or a bonobo. Some people will recoil at this idea. The whole concept of an ape is so emotionally loaded. But let's examine one other characteristic of our makeup that can be as strong an influence on our behavior as the models and scripts we use.

Bonobos used to be called pygmy chimpanzees but are now recognized as a separate species of ape. Humans are equidistant from chimps and bonobos in terms of genetic makeup. (The historian Jared Diamond suggests that we should be considered a third species of chimpanzee.) Those apes are our closest relatives and share more than 98 percent of our DNA. We each, in effect, have one of these apes inside of us. Its ability to take over our behavior depends on many variables, such as our genetic predisposition toward self-control. Though the frontal cortex puts the brakes on the emotions, our upbringing will influence our ability to control emotions, as will the temporary social context in which we find ourselves. But it seems clear that the inner ape has at least some power over all of us some of the time and under the right circumstances.

A startling experiment was carried out by Philip Zimbardo in the psychology department at Stanford University in the summer of 1971. The twenty-four participants in the experiment were given psychological tests to make sure they weren't sociopaths or serial killers. In fact, they were all college kids from the area around Palo Alto. In Zimbardo's words, they were "an average group of healthy, intelligent, middle-class males." The researchers built prison cells in the basement of the psych building on campus. Then Zimbardo assigned half the group to be guards and half to be prisoners. He tried to make it as realistic as possible. He gave the guards sunglasses, and they wore uniforms. While they understood that it was just an experiment, the prisoners were actually arrested in their homes by real police with fake warrants and taken in handcuffs to the real Palo Alto Police Department. They were fingerprinted and blindfolded. They were dressed in prison jumpsuits.

Despite everyone being aware that it was an experiment, within a matter of days the guards had become so abusive and the prisoners so recalcitrant that the experiment had to be shut down. These peace-loving, educated, well-bred participants moved so rapidly back in evolutionary time that they had, in the words of one of them, created an "atmosphere of terror" within forty-eight hours.

After the scandal at Abu Ghraib prison broke in 2004, Zimbardo published an article that made it clear that those who committed the acts of torture in Iraq were not "bad apples." They were no different from the rest of us. Referring to his original experiments in 1971, Zimbardo wrote, "The planned two-week study was terminated after only six days because it was out of control." Zimbardo chose the subjects for the experiment specifically for their qualities of normalcy and good mental health. Nevertheless, "The terrible things my guards did to their prisoners were comparable to the horrors inflicted on the Iraqi detainees." Even the details were similar, such as hooding the prisoners, soaking them with hoses, and forcing the men to mimic having sex with one another.

Under the right circumstances and to varying degrees, any of us can quickly begin to behave like chimpanzees, and presumably like *Homo erectus* and the australopithecines once behaved. That proclivity toward slipping back in evolutionary time into what we could call traditional ape behaviors can help explain much about our world, such as why people kill one another en masse. The answer is: because we can. With our large brains and clever hands, we're able to do far more damage than other apes. The list of places where genocide has taken (or is taking) place includes every part of the earth where humans have ever resided. There are well-respected researchers, Diamond among them, who believe that we are the only remaining species of *Homo* because we killed the others off.

Zimbardo said that during the experiment in 1971, he was completely unprepared for how quickly the transformation took place in his students. It wasn't as if they had to learn a whole new set of mental models and rehearse new scripts for their behavior. Ancient ones were already there, just waiting to be activated. Pinker writes about how certain information is innate in us, what he calls "standard equipment," much of it vestigial. "People hold many beliefs that are at odds with their experience but were true in the environment in which we evolved, and they pursue goals that subvert their own well-being but were adaptive in that environment." One of those vestigial pieces of standard equipment appears to be a capacity to delight in the torment of others. He goes on:

> *Homo sapiens* is a nasty business. Recorded history from the Bible to the present is a story of murder, rape, and war, and honest ethnography shows that foraging peoples, like the rest of us, are more savage than noble. The !Kung San of the Kalahari Desert are often held out as a relatively peaceful people, and so they are, compared with other foragers: their murder rate is only as high as Detroit's.

In other words, the Garden of Eden described in Genesis is a bit of fanciful wishful thinking on the part of the authors. Go back as far as you like. You will not find it.

I know one bonobo at the Milwaukee Zoo who is old and blind, and some of the others lead her around and care for her. The fossil record shows people who were so old that they had no teeth and could not walk. The assumption is that others cared for them. So I'm confident that we have long been capable of the most admirable sort of human generosity and kindness. But at some level, we're also capable of the depths of depravity. It's significant that Zimbardo's prisoners and guards were explicitly defined as two distinct groups. And as we'll see, this has important implications for how we behave today.

Most of us will never be prison guards or engage in (or be victims of) genocide. But many of us will at some point have the experience of behaving in ways that shock us, as if someone else did the behaving while we watched helplessly from the sidelines. And some of us may see a complete split in our behavior between what we think of as our true selves and another creature that seems foreign to us. What we call civilization has existed for a fraction of a second in evolutionary time. Lucy's kind lived for 50,000 generations, about a million years. Our kind have lived for 5,000 generations. We've had agriculture for about 500 generations, and cities are even more recent. While social customs, culture, and technology have been changing at breakneck speed, our brains and bodies have evolved slowly. As Johanson put it, "Biologically we are, in many ways, still foraging the savannas, but culturally we are exploring outer space."

Daniel Dennett, codirector of the Center for Cognitive Studies at Tufts University, writes that "Every bargain in nature has its rationale. . . . But a rationale can become obsolete. As the opportunities and perils in the environment change, a good bargain can lapse." Evolution continues, but we haven't had enough time to change that much during the few thousand years since we were all hunters and gatherers, when our only job was survival.

In my quest to understand a bit more about human behavior, I stumbled upon the story of John Tanner. He was born around 1780 near the Kentucky River and was kidnapped by Shawnee Indians when he was nine years old. The Shawnee sold him to an Ojibwa family. Tanner was given the name Shaw-shaw-wa-be-na-se, which means "the falcon." He was brought up in a traditional hunter-gatherer society that ranged more than 1,000 miles from Michigan and Minnesota up to Hudson's Bay. Like the Jarawa, who survived the tsunami on the islands of the Indian Ocean, the highest native technology of Tanner's kidnappers was the bow and arrow. They had no fixed dwellings but lived in temporary wigwams. They had scant agriculture. Like other apes, they lived in traditional groups: overnight groups of twenty to fifty people, the cast of characters changing constantly; clans of about 150 people; and tribes of a few thousand who shared a language. Although they had been given some modern technology, such as guns, by the European fur merchants with whom they traded, in many ways they lived a life like the one humans would have lived beyond the horizon of 11,000 years ago, before which agriculture was uncommon. What makes Tanner's story unique is that by the age of nine, he'd had just enough experience with the modern world that he was able to act as an observer of the native culture and not just a participant in it. He retained some detachment throughout his life and later returned to recount his experiences in the ways of ancient man in great detail. *A Narrative of the Captivity and Adventures of John Tanner* was published in 1830.

He describes his early days with the Indians much as Viaje might describe his early days among the bonobos of the Milwaukee Zoo, if he could: "When we returned from hunting, I carried on my back a large pack of dried meat all the way to the village, but though I was almost starved, I dared not touch a morsel of it." And: "By one or the other of them I was beaten almost every day." Before his first season was out, one member of the group had tried unsuccessfully to kill Tanner with a tomahawk.

After two years, the Ojibwa sold him to Net-no-kwa, an old woman who was chief of the Ottawa and had lost her own son of about Tanner's age. Here, for the first time since his abduction, he was treated with a measure of kindness and given food to eat and clothes to wear. Moreover, Net-no-kwa was married to Taw-ga-we-ninne, an Ojibwa who treated Tanner with kindness. His stepfather allowed Tanner to prove himself by shooting a pigeon and then gave him a rifle so that he could learn to hunt.

In his descriptions of the wars that raged among the various tribes, Tanner provides good examples of the cultural underpinnings of genocide, not to mention much more benign forms of human behavior. The German ethologist Irenäus Eibl-Eibesfeldt has said, "To live in groups which demarcate themselves from others is a basic feature of human nature." As demonstrated in the Stanford prison experiment, anytime two groups are formed by whatever means, the likelihood is that the interactions between them will become hostile. An experiment called the Robbers Cave Study showed this almost two decades before the Stanford prison experiment. And this raises the question that I will ask repeatedly: Are we doomed to do what's natural? Or can we go against natural law and do better?

Muzafer Sherif, a social psychologist, brought together twenty-two white, fifth-grade boys, all eleven years old. He took them to a camp in Robbers Cave State Park in Oklahoma, while he and his colleagues pretended to be camp counselors. All the boys were doing well in school. All had two parents living at home. None of them knew the others beforehand. Sherif arranged to have one group arrive ahead of the others. They weren't told that there would be another group arriving and, indeed, didn't notice for a time, due to the remoteness of the location.

The first group, with no prompting, immediately began setting up cultural boundaries to define itself as an in-group. They decided to call themselves the Rattlers and stenciled the name on their shirts. Meanwhile, the other group, unknown to and unaware of them, had done the same thing, establishing group norms and calling themselves the Eagles. As described in David Berreby's book *Us and Them*, the two groups separately "created 'our' ways of doing things. Each group settled on a favorite swimming spot, its preferred route to it, its style of dealing with scrapes and sprains."

On the sixth day, the two groups discovered each other, and the first response of both was hostility. The Eagles began referring to the Rattlers as "those nigger campers." Remember, this was 1954 in rural Oklahoma. "Communist" was another epithet flung around freely. Once the groups and the hostility were established spontaneously, Sherif stepped in and organized competitions between the two groups in the form of sports, such as baseball and tug-of-war. The Rattlers had made a flag for themselves and arrived waving it. When the Rattlers won the first game, baseball, the Eagles refused to eat with them. When the Rattlers won the tug-of-war, too, the Eagles retaliated by stealing their flag and burning it. A war was on. It escalated to the destruction of both camps, theft of anything not nailed down, and it required the experimenters to step in more than once to prevent injuries.

Throughout this exercise, the two groups developed differences along cultural lines to distinguish themselves from each other. For example, the Rattlers swore, while the Eagles refrained from swearing. Tanner points out that the Ojibwa and Assinneboins despised each other and waged war and took scalps. The hatred was explained by the fact that one group roasted food on spits, while the other cooked with hot stones.* But the truth appears to be the other way around. Groups first develop hostilities, then invent cultural traits to distinguish themselves or use existing differences to explain the hostilities. Hostility to anyone outside the group, however it's defined, is the norm for creatures of our kind. The Greeks called everyone who was not Greek by the name "barbarian." The fact that we all still carry this potential behavior within us is one of Dennett's "good bargains" that has lapsed, while, as he put it, "it takes time for evolution to 'recognize' this." Patriotism is an example of this, and wars fought behind national flags are a vestigial behavior.

It may seem like we in our modern cities are so far from the Laetoli family, from *Homo erectus*, even from John Tanner's family, that these effects, while curious, really have no bearing on modern life. Neither do subjects in psychological experiments move our hearts. After all, we are thinking creatures. We aren't held in the thrall of ancient instincts anymore. We can make decisions and act accordingly. Well, maybe we can and maybe we can't.

There's an effect that Judith Rich Harris, author of *No Two Alike*, calls "groupness," which embodies the intergroup hostilities and

*The word "Ojibwa" means "roasted until puckered." While most dictionaries explain that this is because of the puckered stitching on moccasins, Tanner's account raises some doubt about that.

xenophobia I've been discussing. The word "groupness" was originally coined by Henri Tajfel, a social psychologist who did experiments at the University of Bristol to show how little it took to trigger the formation of in-groups among people. Almost any slight difference, no matter how abstract or trivial or even imaginary, will set off the process—and the hostilities. In this passage from the book of Judges (12:5–6), less than a syllable defines groupness:

> And the Gileadites took the passages of Jordan before the Ephraimites: and it was so, that when those Ephraimites which were escaped said, Let me go over; that the men of Gilead said unto him, Art thou an Ephraimite? If he said, Nay: Then said they unto him, Say now Shibboleth: and he said Sibboleth: for he could not frame to pronounce it right. Then they took him, and slew him at the passage of Jordan: and there fell at that time of the Ephraimites forty and two thousand.

Another psychologist at Stanford, David L. Rosenhan, performed an experiment that showed just how persistent the groupness effect can be. Rosenhan enlisted seven completely normal colleagues to go to mental hospitals and present themselves with a complaint of hearing voices. All they said was that they heard the words "empty," "hollow," and "thud." They complained of no other symptoms and displayed no other peculiarities. Seven out of the eight bogus patients were immediately admitted with a diagnosis of schizophrenia, while the eighth was labeled manic-depressive and also admitted. As soon as they were admitted, they told the staff that they had stopped hearing the voices and felt fine. Moreover, they never again showed any sign of abnormal behavior. Nevertheless, the doctors and nurses, already part of a well-defined in-group, continued to regard the researchers as mentally ill. They kept them in the hospital for up to fifty-two days, with an average stay of nineteen days. During that time, the doctors dispensed more than 2,000 doses of major tranquilizers to the bogus patients, despite there being no evidence of any disorder.

The researchers at first attempted to conceal the fact that they were taking notes on their experiences. But they soon noticed that the staff simply ignored them as long as they didn't make trouble. (This made it easy for the team to throw the medications away.) The researchers henceforth took their notes out in the open. As Rosenhan wrote in his 1973 paper, "Once a person is designated abnormal, all of his other behaviors and characteristics are colored by that label. Indeed, that label is so powerful that many of the pseudopatients' normal behaviors were overlooked entirely or profoundly misinterpreted." The fact that the fake patients were taking notes was put in their files as a demonstration of abnormality. Groupness dictated its own conclusions. Rosenhan said, "Given that the patient is in the hospital, he must be psychologically disturbed. And given that he is disturbed, continuous writing must be behavioral manifestation of that disturbance." (Curiously, the real patients immediately recognized the researchers as frauds. Their own groupness tipped them off that these were outsiders.)

The drive to fit in with a group is almost unbelievably powerful. Numerous experiments have been performed to show that people will deny the evidence of their senses and even risk their lives to do what others around them are doing. In the 1950s, Solomon Asch, a pioneer of social psychology, performed an experiment that is legendary in academia. A group of people in a room was asked to judge which of the three lines on the right matches the one on the left.

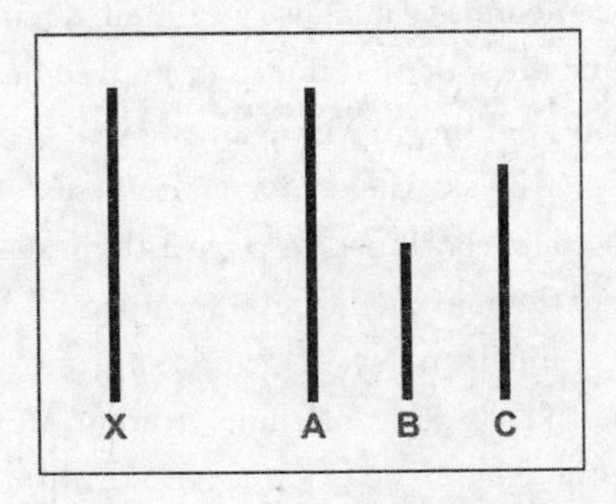

The answer is obvious, but all of the people in the room except one had actually been recruited by Asch to give the wrong answer. More than a third of the test subjects went against their own perceptions in order to go along with the crowd. In another experiment, the subjects were asked to fill out a form in a room full of people. Again, only one person was not in on the experiment. During the session, smoke began pouring out of a heating vent. When everyone ignored it, so did the test subjects, despite the obvious sign of danger.

Groupness, then, is a force that should not be discounted as an influence over our behavior. "The strong emotions associated with groupness were inherited from [our] ancestors," Harris wrote. "They served the same purpose, and were passed down in the same way, as the instinct that impels a bee to give up its life to defend the hive."

As long as we're talking about *Homo erectus* or Jarawa natives or even John Tanner, we find it fairly easy to accept that the influence of groupness has real control over the lives of those people. But as soon as we move into our own highly technical culture, we stop believing. Groupness, for example, would have played an important role in the way people behaved during the tsunami in the Indian Ocean and during the eruption of Mount St. Helens. You don't want to be screaming and running from the scene when everyone else seems to be having fun on vacation. So maybe in those settings we can appreciate how people might be laboring under ancient irrational influences. On the other hand, we might balk at the idea that NASA could be influenced by such forces. NASA, after all, represents the epitome of a rational technical culture, the polar opposite of the behavior we'd expect from a traditional person running naked in the forest. And yet it has repeatedly shown that its people, along with the organization itself, are not immune to the influences that guided our ancient ancestors.

After the space shuttle *Columbia* broke up in flight on February 1, 2003, a commission was formed to investigate the accident. It covered all the mechanical and physical facts of the explosion, but that left a very basic and vexing problem. The accident had been avoidable and had not been avoided. That meant, in effect, that the smartest guys in the world had done the dumbest thing in the world. Twice. The commission's report sought to explain how this could happen by talking about the very sorts of mental models and scripts, along with group hostilities, that shape the lives of the Rattlers and Eagles of this world and that shaped the experiences of John Tanner and *Homo erectus* as well.

Managers at NASA had fashioned a psychological framework that allowed them to systematically ignore clear evidence that they were heading into trouble. NASA's triumphant experiences in putting men on the moon during the Apollo program of the 1960s had led to the formation of mental models and behavioral scripts in the organizational culture that persisted despite drastic changes in the environment, such as greatly reduced budgets and overwhelming evidence that essential pieces of equipment were malfunctioning.

The second force influencing critical decisions at NASA was groupness. The final report of the commission on the *Columbia* accident said, "External criticism and doubt . . . reinforced the will to 'impose the party line vision on the environment, not to reconsider it. . . .' This in turn led to 'flawed decision making, self deception, introversion and diminished curiosity about the world outside the perfect place.' " The report is quoting Garry D. Brewer, a professor of organizational behavior at Yale University, who was attempting to explain how management at NASA could have behaved the way it did. The "external criticism and doubt" came, for example, after the explosion of the space shuttle *Challenger* in 1986, the first time that NASA made the worst mistake it could have made. That criticism came, significantly, from outside of the in-group.*

*Groupness was also an influence in the crash of American Airlines flight 587, which I discussed in Chapter 2. Sten Molin's job had been protected by it. Although he twice mishandled the controls of airliners, he was not relieved of duty, because he was part of an in-group that protects its members.

The combination of groupness and persistent mental models made for an organization that could not take in new information when that information did not accord with its indelible concept of itself as the "perfect place," as Brewer called it. Moreover, it could take any contradictory information and reinterpret it as confirming the existing model. This came about through two major influences that made NASA's models unassailable and made its culture hostile to all outside groups. To begin with, there was the unprecedented investment not just of money but also of personal and emotional effort during the Apollo program. The divorce rate was high, as marriages fell apart under the strain. People literally gave their lives for the effort. Gus Grissom, Roger Chaffee, and Ed White died in a fire during a test on the launch pad in 1967. But following all that sacrifice was an astounding success, arguably the highest achievement of human technical culture, with men walking on the moon while we watched them on television. This has all the ingredients necessary to form robust models and scripts (big investment, big reward). And that experience simply hardened the shell of groupness that already characterized NASA.

NASA's unspoken and unconscious attitude by that time was: We must be right; after all, we put a man on the moon. There had been many reinforcing steps along the way, too. During Apollo 13, for example, the concept that "failure is not an option" was developed, and the safe return of Apollo 13 served to strengthen the models and the ability of groupness to repel ideas from outside. It also promoted a dangerously wrong idea. For failure, unfortunately, is always an option.

If individual human beings can form forceful and persistent mental models, organizations or groups of people seem to be able to do so on an almost unimaginable scale. A person has secret doubts and fears. An organization has the emotional life of a reptile. The *Columbia* accident was essentially a repeat of the *Challenger* accident, not in mechanical terms but in terms of the cognitive mistakes that lay at the heart of the accidents. As the investigating board put it, "Both *Columbia* and *Challenger* were lost also because of the failure of NASA's organizational system. . . . Both accidents were 'failures of foresight' in which history played a prominent role."

In the case of *Challenger*, engineers were faced with the fact that fuel in the solid rocket boosters was burning through the rubber O-rings that sealed the seams where sections of the rockets were joined. Groupness dictated that no one outside that immediate culture was fit to judge the fruits of their labors. Confirmation bias is a phenomenon in psychology by which people tend to take any information as confirmation of what they already believe. In addition, they tend to ignore or miss any information that doesn't confirm what they already believe. This can work to gradually revise a mental model in a one-way direction. Because NASA believed that "we're the best" and that "failure is not an option," all information tended to support that conclusion, no matter how contrary it might have seemed to an outsider.

Each time the solid rocket fuel burned the rubber O-rings during launch without an accident happening, the engineers at NASA readjusted their models and scripts slightly to accommodate that as "normal." Through a subtle progression, a complete failure of design was turned into an acceptable situation. Each time nothing bad happened, they did it again. This confirmed the mental model, even while groupness helped to keep conflicting information from having any effect. The weather was the spinning roulette wheel in this complex system that NASA managers had unwittingly set up for themselves. All they needed was for cold enough weather to coincide with a launch, because cold made the rubber O-rings more brittle and therefore more likely to burn through. It was just a matter of time. The ape-like hierarchy at NASA ensured that those engineers who knew or suspected the truth would not be heard. Like Viaje, the lowly new member of the bonobo group in the Milwaukee Zoo, those engineers knew better than to speak up about potential problems.

The alpha individuals were running things and punished dissent harshly. Like Viaje, those engineers knew that it just wasn't worth the beating.

The same array of troubles bedeviled *Columbia*. Insulating foam blew off the main fuel tank and hit the orbiter. The engineers had seen it happen a number of times, but management kept on launching anyway. When nothing bad happened, they took that as confirmation that they were right and reset their mental models to accommodate the malfunction. As the final report on the accident clearly stated, "The initial Shuttle design predicted neither foam debris problems nor poor sealing action of the Solid Rocket Booster joints. To experience either on a mission was a violation of design specifications. The anomalies were signals of potential danger, not something to be tolerated."

But in the culture that had evolved at NASA, each return from a successful mission was another moon landing. If the world had largely come to ignore space launches, NASA was still hearing applause that was, by the time of *Columbia*, more than thirty years old. So, instead of peering more deeply into the problem, they gradually revised their models until they were literally interpreting failure as success. The final report of the commission said:

> Engineers and managers incorporated worsening anomalies into the engineering experience base, which functioned as an elastic waistband, expanding to hold larger deviations from the original design. Anomalies that did not lead to catastrophic failure were treated as a source of valid engineering data that justified further flights.

This is essentially the same mistake the students at Texas A & M University made when they increased the size of the bonfire year after year. Every time it didn't collapse, that validated the model for increasing the size the next year. And this is precisely how a mental model can be expected to function. It operates on a simple rule: if nothing bad happens, you must be doing something right. So influential were NASA's models and scripts, and so delusional its self-confidence bred of groupness, that even after *Columbia* broke up, killing all on board, the space shuttle program manager told the press that he was "comfortable" with his previous assessments of risk and didn't think the foam debris had caused the accident. But remember that a key feature of this system is that, taken one small step at a time, each decision always seems correct.

It might seem at times that the behavior of apes and our ancient ancestors has little to do with our modern life. But Lucy, the Laetoli family, *Homo erectus*, John Tanner, and the soon-to-be-extinct Jarawa are all operating in the background of every decision we make. It seems to me the height of good sense to see ourselves reflected in the ways of ancient people, at least so that we might do a bit better than they did.

NASA's behavior, like that of the students at Texas A & M, provides a model for our collective behavior as a society. Each of the behaviors that have wound up contributing to our predicament has its origins in a large investment of effort that produced triumphant results. Despite signs of trouble, we adjust our mental models to accommodate larger deviations from the norm. Our groupness helps to keep bad news from upsetting our view. Without a mechanism for reframing our behavior or redefining our group, the effects are ignored, as they were at NASA, until a catastrophe happens.

Six

The Corporate Emotional System

It sometimes appears that one of the underlying principles with which we are forced to live is the principle of unintended consequences. Throughout our history we human beings have been very good at striving to achieve whatever ends we can imagine, whether they be ensuring a steady supply of food, conquering a neighbor, or curing a disease. But there are costs associated with our successes, and they often seem to come out of nowhere, as if nature were deliberately penalizing us for our very cleverness. No good deed goes unpunished. No bad deed, either. No deed at all, it seems. Everything has a cost. (As we shall see, this concept—the universality of costs—is reflected in fundamental laws of nature.)

A remarkable group of people came together in Cambridge, Massachusetts, in 1949. Or perhaps anyone who had been put into that class at that moment in history would have turned out remarkably. Their story is a tale of unintended consequences. Most of the members of this group could not have attended Harvard Business School if the GI Bill had not been paying their tuition. They were, by and large, people of modest means who had endured the Depression and fought in World War Two. They shared a belief in living through hard work and sacrifice. During the decades to come, 28 percent of the 700 members of that class became the most successful business executives in American history. Their careers, which are now over, represent a development in American business and culture that will likely never be seen again. Their efforts brought to fruition the consumer culture that had been on its way when they were children but had been interrupted by the Depression and the war. During the 1950s and 1960s, they redefined what industry did, inventing the new technologies that have proliferated everywhere. They also redefined the way Wall Street worked, vastly expanding the ability of corporations and individuals to make money. The way of life I see around me right now is part of their legacy, from hybrid cars and instant satellite images of weather on my laptop computer to heart transplants and hundred-channel television.

Within a decade of their graduation, many of the members of that class had begun to see that the way forward in American business was going to be inextricably bound up with technical innovation. The industrial age was over. The fields of nuclear physics, material science, aeronautics, medical science, computers, lasers, and chemistry were all advancing. The launch of the satellite *Sputnik* in 1957 put the United States on notice that the Russians had developed rockets powerful enough to deliver atomic bombs to our cities. The development of technology was instantly turned into a fight for survival and, as a result, there was a tremendous churning of imagination, invention, and manufacture. Basic science was put in the service of practical technology on a scale never before seen. New products were brought forth, old ones were transformed, and the business of making and selling things itself was revolutionized and rebuilt according to the modern blueprint for efficiency, marketing, and military might.

The wealth that was produced as a result of those efforts was unmatched in history. It completely changed the face of America—

and is now in the process of changing the entire world. Such radically different achievements of our society as the exploration of space and the building of the interstate highway system—the largest public works project in history—were made possible by that sudden and unprecedented creation of wealth. I was born in 1947, and this creative upheaval formed the backdrop of my life. The people who made it happen were my father's age. This new culture gradually assembled itself around me as I grew up. It seemed like magic to me, as if I'd been born a pauper and all at once a castle began growing up around me until at last I was transformed into a prince.

But I remember a time before all this opulence. When my father received his PhD in biophysics in 1953, we drove from St. Louis to his first job in Houston, Texas. There were no interstate highways. It was two-lane blacktop all the way, and in places the old road had been torn up and we drove on bare dirt. My mother tied wet handkerchiefs around our faces against the dust. The car had no air-conditioning, no radio. It was a different culture. Everybody smoked cigarettes. Children were dying of polio. There was no McDonald's, no Holiday Inn, no FedEx, no Xerox. But there was about to be all that and more.

C. Peter McColough was a member of that class of 1949 at Harvard Business School, and his professional trajectory tracks the complete transformation of the American social and economic landscape. His company, Xerox, was intimately involved in it. The tale of how Xerox nearly lost its life as a result of decisions McColough made as CEO has been recounted numerous times. But it's worth examining as a story of mental models and scripts, of groupness, and of behaviors rooted in our ape ancestry. It is a story that is uniquely American. It is also representative of the achievements of the Harvard Business School class of 1949 and of the unintended consequences and costs of those achievements. It shows how what we think we are doing and the outcomes of our actions are often strangers to each other.

Around the time that my family and I were driving the dirt roads from St. Louis to Houston, Peter McCologh was a young man interviewing for a job with the Haloid Company in Rochester, New York. Over the next decade or so, he helped to transform it into the monopolistic giant known as Xerox. The Haloid Company could not have been more obscure. It made an odd industrial copier that used static electricity to apply a fine black powder to white paper. It heated the paper until the powder melted, kind of like melting plastic, and then immediately cooled it again so that the powder solidified. Even when fifteen years of development had refined it for office use, the awkward contraption weighed 650 pounds and took up enough space to require its own room. How could such a Rube Goldberg device be made into a commercial success? IBM took a look at it and concluded that it would never make it to market.

Moreover, it broke down constantly. Before the first copiers were delivered to customers, McColough had to design and bring to life a national sales and service force to tend and mend the balky copiers. Who could have guessed that businesses all over the country would welcome those behemoths into their coffee rooms? Who could have anticipated our insatiable appetite for copies of copies of copies of documents? I am drawn to the story of Xerox because it's so American. If the story of DNA is a tale of urgent reproduction, then in America we see its natural outcome: mass production. Xerox elevated this concept to an unprecedented level of purity and abstraction. Xerox is the story of endless duplication.

There were plenty of copying machines around at the time, but they were sloppy, tedious, and produced ugly copies. The idea driving Xerox was to make that process more elegant and convenient. Without McColough's genius for marketing, though, all the effort might have come to nothing. The machine cost $30,000 to buy, compared with around $400 for the nicest wet-process copier on the market at the time. (The word "xerography" is from the Greek for "dry writing.") McColough devised a scheme by which a company could lease the machine for a flat fee of $95 that included the first 2,000 copies. The copier had a meter that kept track of additional copies, for which the customer was billed. The people who worked in the offices of America went copy-crazy. Xerox originally estimated that each machine would turn out 10,000 copies a month. The real figure was closer to 40,000. During the fifteen-day trial period, almost any office could become hooked on the push-button copier. *Fortune* magazine called the 914 copier "the most successful product ever marketed in America." Xerox held exclusive patents on the technology. It was as if the company could just . . . Xerox money. Sales went from $40 million in 1960 to $3 billion* just over a decade later. In one year alone, 1962, sales increased by 70 percent. By the time McColough took over as president in 1966, he was one of the richest graduates of the 1949 class at Harvard Business School.

McColough had every reason to be under the influence of some very stable mental models and a very strong sense of groupness within Xerox. By definition, the people at Xerox were winners. By definition, they knew what they were doing. McColough had endured the Depression, won the war, and made his fortune. He was a well-liked, energetic, and impatient man with ample evidence that if he just made up his mind and worked hard, he and his company could achieve whatever they wanted. This is exactly why success is such a dangerous element of any endeavor. Embrace the struggle. Beware the achievement. For it steals your caution even as it leads you down the next unknown pathway.

McColough shared with his mentor and father figure, Joe Wilson, the former CEO, the vision that Xerox would dominate the digital revolution that they correctly saw coming. When Wilson appointed

*It's easy to lose sight of the size of such numbers, because they're so casually tossed around. Translating numbers into time can give us a better sense of their magnitude. A million seconds, for example, is 11.6 days. A billion seconds is 31.7 years.

McColough CEO in 1968, the youthful McColough, at age forty-six, was determined to see that vision become a reality. The world of computers at that time consisted of what people in the business called Snow White and the Seven Dwarfs. IBM was by far the leader, while seven smaller companies competed for most of the remaining business. McColough reasoned that with the amount of money he had at his disposal, he could simply buy one of those computer companies as a way of jump-starting the digital revolution at Xerox.

But both he and Wilson were operating on a faulty mental model. It was as wrong at its core as was NASA's model of itself as the perfect place. It said that Xerox was the best technology company, because a copier is technology and—well, look our profits. (The 914 copier was the equivalent of their moon landing.) Moreover, the model was based on a fundamentally flawed analogy: copiers were boxes with electrical stuff inside of them; so were computers; therefore, Xerox was best suited to lead the digital computer revolution.

In addition to the flawed analogy, Xerox had another problem: the 914 copier, from which all the money flowed, was a fluke. Profitable as it might have been, it was a unique invention. New models would follow, but no new concept. And only the patent gave Xerox a monopoly. Without something new and original to follow the 914, Xerox was vulnerable to competition. The fact that the 914 was an analog fluke only made matters worse. Xerox knew nothing about digital computers. Both Wilson and McColough understood this, which only added to the urgency McColough felt to act.

Moreover, the market for copiers was about to be overrun with competing devices. Both Kodak and IBM were hard at work developing copiers, and a consortium of Japanese firms was determined to capture the American market as well. The nature of the copier business was changing rapidly, but the shared mental model of Xerox as the best technical company caused those in charge to discount such developments and plunge ahead. "Xerox" had become an everyday verb. How could they go wrong?

There was, indeed, a digital revolution going on, and none of the players wanted to relinquish its spot for fear of being left without a seat when the music stopped. McColough's attempts to buy one of the Seven Dwarfs computer companies were all rebuffed. Smarting from that experience of failure, McColough was emotionally primed. That made him more likely to act. All his experiences in life had taught him this lesson: effort equals success. If at first you don't succeed, try, try again. And that forceful action turned out to be exactly the wrong sort of lesson to learn when what was needed was new learning and quiet reflection.

The trouble with our emotional systems is how smart they are. They learn really well. In ancient times, success was a fairly simple matter. For example, forceful and persistent effort at hunting would yield food. If the emotional system learned that as a favored behavioral script, it would probably serve to keep us alive. But in our modern world, things are not so simple. And forceful concerted effort may be good in one situation and simply out of place in another. Lao Tzu wrote in the *Tao Te Ching*:

Success is as dangerous as failure,
and we are often our own worst enemy.

One situation may look like another and be completely different, fooling us into pursuing the worst possible strategy based on the best possible evidence from our emotional system: our past success.

Certainly there were other factors involved, but the influences I've been describing were essential features of the landscape, both internal and external, that McColough faced as he planned his next move in the late 1960s. McColough was known to be headstrong and impatient. He was emotionally primed from being repeatedly rebuffed. He correctly understood the costs of being left out of the computer revolution. And he was determined to be right. One of the few remaining American computer makers was a niche company called Scientific Data Systems. McColough did not even take the time to get his ample staff at Xerox to evaluate the company's books. He met with Max Palevsky, the founder of SDS, briefly and in secret. Then he spent nearly a billion dollars to acquire the company, laboring under the false mental model that said, essentially, all computers are the same. Any port in a storm.

Jack Goldman, who had recently been hired as the chief scientist at Xerox, was shocked that his boss hadn't consulted him on the largest technical acquisition in the company's history. He was equally shocked to learn that even the engineers at Xerox knew nothing about computers. But Goldman was in the same position as the engineers at NASA, who saw the rubber O-rings burn through and desperately wanted to tell their bosses to stop launching until the problem was solved. Or the ones who saw the foam insulation hit the wing of *Columbia* and desperately wanted the military to turn its satellite cameras on the craft to have a closer look at the damage. Those people weren't high enough in the group to do that. They didn't have enough status. And anyway, in Goldman's case, it was already too late.

Down the road in Massachusetts at that time, a group of mad young geniuses were creating the Data General Corporation in a crucible environment that would have scorched the hair off the serene accountants who ran Xerox. Tortured and inspired, those engineers and microcoders spawned by the Digital Equipment Corporation stayed up all night in dank basements, disheveled and scrawny, breaking all the rules and taking huge risks to create inspired technology.

In the real world of the digital revolution, an entire company's fortunes could turn on the brilliance of a single engineer, as Data General's did on Carl Alsing and Digital Equipment Corporation's did on Gordon Bell. Xerox did engineering by committee. Making matters worse, the company that McColough had purchased, SDS, had already matured and was on the wane. In the next three years, it would lose $120 million as its business stagnated.

Success had other unintended consequences at Xerox. Because

the 914 copier took off so rapidly, the company was forced to grow at an abnormal rate, like a cancer. To manage its runaway growth, McColough was forced to hire well-established executives who knew how to manage large organizations. He found them in the auto industry. This betrayed another naïve assumption based on faulty mental models: that Xerox was just another industrial company in the great American tradition.

In fact, Xerox stood at an inflection point in American business. It could have become the technology company that Wilson had envisioned, except that it was still too bound up in the history of American business. Like the Laetoli woman, groping for but unable to see Safetyland, McColough could reach for but not quite envision what the corporation of the future would look like. At the beginning of the 1960s when the captains of industry discussed the world of big business, they were talking about Detroit and the automakers and Pittsburgh with its blast furnaces turning out steel. A hierarchical business structure had served heavy industry and the factory floor well, but it was not the stuff of the future. The new high-tech companies would thrive on a very different structure, one that involved dynamic control and bottom-up development. By hiring executives away from those legacy businesses to run Xerox, McColough essentially hobbled the company with a sluggish bureaucracy that prevented it from taking advantage of its own good fortune. Run by Ford and GM executives, Xerox would never be as nimble as it needed to be to compete with the likes of Data General or, in the end, even the tiny upstart known as Apple. Xerox was caught between a foregone industrial age and an age of computers not yet born. This had one practical consequence that trumped all others. Even if Xerox could build the best computer in the world, it was doomed to fail. As one of the top executives at Xerox put it at the time, the company's bureaucracy resulted in a cost structure that was prohibitive. "If we built a paper clip," he said, "it would cost three thousand bucks." And that's essentially what happened.

Jack Goldman convinced McColough to fund the Palo Alto Research

Center in 1970 to be a laboratory for cutting-edge research akin to Bell Labs at AT&T. As many people know by now, the scientists at PARC invented the personal computer, the mouse, the graphical user interface that we take for granted, Ethernet, laser printing, and almost everything else associated with the digital revolution.

McColough imagined that the engineers at SDS would implement the inventions coming out of PARC. But he had not taken into account the groupness effect. The engineers at SDS were the Eagles, and PARC represented the Rattlers. SDS resented all the money that PARC was receiving to develop what they thought was rightly theirs: computer technology. Like the boys at Robbers Cave, they dug in their heels and refused to cooperate with PARC. Between the resistance at SDS and the lumbering bureaucracy at headquarters, the collaboration never happened. By late 1970, Xerox began missing its earnings forecasts. Before it was all over, Xerox would lose tens of billions of dollars on a side road to nowhere.

PARC truly was a hotbed of innovation, but it wasn't immune to the psychological and emotional ailments of success, either. Brilliant as they were, many of the scientists there were suffering from a severe case of groupness and saw themselves as too sophisticated to get involved with the hobbyist computer culture that had sprung up when cheap microprocessors began flooding the market. But it was precisely there, in that scruffy subculture, that even madder and younger geniuses were doing the same sort of thing that Data General was doing: creating sophisticated, practical computers.

Xerox, to its credit, was building some fine computers by that time. But they housed tens of thousands of dollars in parts alone. The Dorado was one such offering and was considered a breakthrough at a list price of only $50,000 (compared with the competing VAX from Digital Equipment Corporation, which cost half a million). The Dorado used 2,500 watts of electricity, was taller than a man, and had to be supplied with 2,000 cubic feet of air each minute to cool it. At that moment, the kids in California were selling ready-made computers that could do practical office work for a few hundred dollars. When one scientist at Xerox, Larry Tesler, warned the head of PARC that an organization called the Homebrew Computer Club would overtake Xerox, the manager said that was impossible, because PARC already had the smartest people working for it. (This is the "perfect place" syndrome that we saw at NASA.)

Much has been made of how Steve Jobs—a Homebrew Computer Club member—and his engineering team at Apple waltzed into Xerox PARC in December 1979 and stole everything they needed to build the Macintosh and revolutionize the computer industry. It's a much more complicated story than that—too complicated to recount in its entirety here*—but the drama had all the elements of the NASA story. Top brass, in fear of being left out of the computer revolution, had made a deal with Jobs: the ability to buy Apple stock in exchange for a visit to PARC and a demonstration of its latest inventions. (This was before Apple stock went public.) PARC scientists objected strenuously to revealing their secrets to Apple, though they were too low in the hierarchy to stop it. The demonstration that was given to Jobs and his team was conducted by grudging engineers, who later could only say I told you so. Tesler was soon hired by Jobs to help design the Macintosh and ultimately to become chief scientist at Apple. Xerox sold its coveted stake in Apple, thus missing out on the most profitable period of growth for that company. Xerox had been the first company to ride a new technology to sales of a billion dollars in less than ten years. Apple was the second.

By 1980, engineers at IBM were hastily throwing together cheap components while executives met with a kid named Gates. The quick and dirty IBM machine, running Gates's clunky Microsoft operating system, appeared on the market in August 1981. For the first time in its history, profit-sharing was suspended at Xerox.

---

*For a good account of this, see *Dealers of Lightning* by Michael Hiltzik.

The behavior of the people at Xerox reminds me of a group of Scandinavians who settled in Greenland about 1,000 years ago. They liked the look of the land, which resembled their home country, Norway, in superficial ways (kind of the way computers and copiers resemble each other, both being made of electrical stuff).

Based on this flawed analogy, the Greenland Norse immediately set about doing what had worked in the past: farming, herding animals that grazed voraciously, cutting down trees for wood or burning them off for pastureland. They built their houses of turf, which they cut at the rate of about 10 acres per house. This worked for a while, but eventually they ran out of trees, while the grazing and cutting of turf left the land unprotected. The topsoil washed into the sea. Crops began to fail. Without wood, they could no longer make charcoal to smelt metal. Their tools wore out and couldn't be replaced. They had happened to arrive in Greenland during a particularly temperate period, but as the climate returned to its colder norm, the effects of their activities grew more pronounced.

The Inuit lived there, too, but they had long ago adapted to the environment. They used whale blubber, not wood, for fuel. They ate it as well, and it provided a high-density source of calories. They lived in igloos made of snow. They used kayaks made of scant frames of wood or bone covered in sealskin. (The Norse boats were all wood.) The Inuit offered their techniques for survival to the Scandinavians, who, because of groupness, would not consider using them. Jared Diamond describes this in his book *Collapse*. He wrote, "Both the failure to develop trade with the Inuit, and the failure to learn from them, represented . . . huge losses to the Norse, although they themselves evidently didn't see it that way." And: "The Norse starved in the presence of abundant unutilized food resources." The result of this behavior was that the Greenland Norse went extinct. The Inuit are still there.

Our behavior in relation to our own environment is beginning to resemble that of the Greenland Norse to an increasingly alarming degree today. And if we keep it up, not even the Inuit will survive. The whales they hunt have accumulated so much DDT and PCBs in their systems that the Inuit are being poisoned by their traditional source of food.

The combination of mental models, behavioral scripts, and groupness can sometimes seem perfectly designed to destroy not only corporate profit but entire organizations in almost any realm. Quaker Oats's great success with Gatorade caused it to apply a faulty mental model to Snapple. This model held (unconsciously) that all drinks are equal. The in-group concept convinced Quaker marketing executives, flush with victory, that they knew better than anyone how to sell drinks. Indeed, Gatorade, which Quaker bought in 1983, was the company's moon landing. When the company bought Snapple in 1994, it was blinded in much the same way that the leaders at NASA had been. Quaker executives didn't see that Snapple had a following and a distribution system that were completely incompatible with the ecosystem in which Gatorade thrived. Indeed, Gatorade and Snapple were as different as my grandmother's ashtray and a real rattlesnake. While Quaker Oats engaged in a series of attempts to bring Snapple to life within the manufacturing, distribution, and advertising environment of Gatorade, they lost more and more money, even as Lipton, Ocean Spray, Nestea, Coca-Cola, and others were cutting into the market. *Business Week* named the purchase of Snapple one of its top ten worst mergers. Quaker sold Snapple in 1997. Quaker had grabbed the rattlesnake and was bit to the tune of a billion dollars. Viewed another way, Quaker attempted to colonize a land that looked hospitable but was not.

At the Schwinn Bicycle Company, a marketing executive expressed the delusional power of corporate groupness succinctly when he said, "We don't have competition; we're Schwinn." The company was subsequently wiped out by mountain bikes, racing bikes, and a Chinese bicycle company that Schwinn created when it moved its own manufacturing offshore.

Even Enron is an example of the influences I'm discussing. While the failure of that company is widely seen as a criminal conspiracy—and was, of course—it came to that pass by a process so gradual that I'm willing to believe that at least some of its principals weren't even aware that the transformation had taken place. Like the engineers at NASA, they developed an elastic waistband for larger and larger deviations from good practice. They shifted their mental models each time something bad happened, and pretty soon the executives were accepting as normal what was in fact illegal, without even being able to remember when they'd crossed the line. Sound advice was rebuffed by groupness or stifled by the social hierarchy.

Because of the natural and inherited way that we process information, this tendency to be overwhelmed by change in the environment has shaped human affairs throughout history. Not long after the last of the Greenland Norse died out, Francisco Pizarro came to Peru to confront the Incas, who were led by an emperor named Atahuallpa. Pizarro's men numbered fewer than 200, while Atahuallpa had 80,000 troops defending him. Nevertheless, Atahuallpa had numerous established models and scripts that allowed him to walk into a trap and be captured. His army had just won several important battles, giving him great confidence. His groupness led him to badly underestimate the Spanish. He looked down on them as uncivilized brutes without sophistication or manners.

When he went to meet with Pizarro on November 16, 1532, he arrived accompanied by several thousand Incas, many of them armed and armored. Although he had experienced civil war, he had never heard of someone from elsewhere invading his land with the intent of conquering his empire. Based on his available heuristics, such a thing was inconceivable, even in principle. Moreover, he had never seen or heard of horses, steel weaponry, or guns. Nonetheless, his attitude was that he knew his world and the forces in it. Pizarro's troops killed 7,000 of Atahuallpa's men the first day, executed Atahuallpa, and within a few years had, for all intents and purposes, brought the Incan empire to an end.

Andy Grove, the cofounder of Intel, has described how difficult doing battle against these natural systems can be. Intel, which was founded to manufacture memory chips for computers, had been gradually overwhelmed by low-cost Japanese copies. By the early 1980s, Intel's memory business was losing money. Grove knew—everyone knew—that memory would henceforth be a commodity. Intel was a memory company. But the company would never be able to compete with Japan on price. What was Grove to do?

It turns out that engineers at Intel had invented the microprocessor chip in 1971. This novel "computer-on-a-chip" could still be sold at a profit, and other companies would be a long time catching up. On the other hand, at the top of the company, where Grove stood vacillating, as he put it, "We had lost our bearings. We were wandering in the valley of death." This is how deeply embedded these forces can become. Even knowing the facts, they found it almost impossible to go against the models and scripts and the cultural definitions that establish groupness, such as: we are a memory company. The change to making microprocessor chips would be the equivalent of the Greenland Norse agreeing to eat blubber.

More often than not, because of its role in our survival, the emotional system gets its way. It does its work without asking permission and for the most part without even making us aware of what it's doing. Some of the most important decisions in life are made this way. Our thinking brain is often a sideshow, because if the emotional system fails to do its job and react automatically to ensure our survival, then the body is in danger. The evolutionary logic is unassailable: What good does reason do if you're dead?

When Grove talks about how and why CEOs are replaced, he

admits that the new CEOs "have only one advantage, but it may be crucial: unlike the person who has devoted his entire life to the company . . . the new managers come unencumbered by such emotional involvement. . . ." In effect, he's saying that the directors on the board don't really want to get rid of the old CEO, they just want to fire his emotional system. They want to revise his mental models and behavioral scripts. They want a new kind of groupness. Successful corporate change initiatives are really about creating new models and a new definition of groupness. And that's not merely difficult to do in the corporate setting, it can sometimes be close to impossible.

But Grove and Gordon Moore, who was the CEO of Intel at the time, succeeded in mid-1985. They made the monumental effort to erase their own scripts and models and group identity and rebuild them. They sat down and asked themselves: If we were fired, what would the new leadership at Intel do? Then, as Grove put it, they decided to "walk out the door, come back, and do it ourselves." Grove later described this experience as one of actually losing his identity. He was right: the emotional system and its trappings are what create the self. They work like an immune system to define inside and outside. In a sense, there is no way around groupness because of that. Success is self-limiting. The only hope lies in changing the group identity when necessary to better suit the environment. Then, with luck, the new group can be successful, too.

After wandering for a year in his "valley of death," Grove made that hard decision to get the company out of memory and devote all of Intel's resources to microprocessors. The company spent another two years recovering. The experience exacted a huge toll, which may be one of the reasons such decisions are so hard to make: the emotional system does what rewards it, not what punishes it. But one big difference between being human and being some other animal is that we can consciously decide to do what is painful in order to achieve a benefit in the future. We can put off going straight for the reward that is most obvious and most immediate. We can step to the side instead, so to speak, and take a less obvious path that will eventually lead to greater reward. I'll discuss this later in terms of the natural laws that allow nonliving systems (weather systems, for example) to resist taking the straightest, most obvious path. They, too, can step sideways at times to reach a better path. As we'll see, even nonliving systems like storms seem to create memories that work like mental models.

But whether in human decision-making or in natural systems, there is always a cost involved in these sidestepping paths. The decision that Grove and Moore made resulted in Intel losing more than $180 million in 1986. The company had to close factories and fire 8,000 employees. But the result was one of the most successful companies to come out of the high-tech revolution. The computer sitting on your desk or in your lap almost surely contains one of Grove and Moore's microprocessors.

As those two pioneers of the computer age proved, the brain is remarkably plastic. Who we are and how we define our group are both malleable constructs, formed in response to the demands of our environment. Because the environment is always changing, we grow rigid at our own peril. It's significant that Grove had seen the worst his environment had to offer: he had escaped the Nazis and Communists and lived through the Hungarian Revolution. He'd come to know firsthand how fickle and deadly an environment can be and how flexibility can save you. He'd even changed his name three times to survive. He had never really dropped his guard.

Seven

The Teachings of Don Juan

One of the joys of my work is that I get to travel to interesting places in my quest to free myself from ignorance, prejudice, and superstition. So it was that I decided to find a place where there was no vacation state of mind, where people had not yet dropped their guard. I considered the Indian Ocean. That's where the Jarawa tribe survived the tsunami. But that seemed somehow too remote, not so much geographically as culturally. It was hard to imagine myself running naked in the forest with bow and arrow. I went instead in search of people who were much more like my own ancestors, in the state of Chihuahua in Mexico.

People began heading south from somewhere in Asia about 20,000 years ago and reached Mexico perhaps 7,000 years later to mingle with elephants, lions, cheetahs, and the other great mammals that lived there at the time. It was the place where my father's mother's mother's mother was born. In all likelihood, some of my own relatives had lived in caves there, perhaps as recently as the nineteenth century. Much like our belief in the power of reason, we harbor a belief that we have tamed our world and know it well, and that where we don't know it, our inventions will serve us to dominate it. Ask the Greenland Norse: the precincts of our real control are fragile islands of illusion. The region I chose to visit is only 350 miles from El Paso, Texas. But it contains the most uncharted territory between the Arctic and the Amazon.

My daughter Elena accompanied me. The journey took us two days of flying and driving before we reached the town of Batopilas at the bottom of Copper Canyon. Our guide was Carl Franz, coauthor of *The People's Guide to Mexico*, who has lived and worked in Mexico for most of his life. Our driver, Chuy (short for Jesús), liked to keep his .45 on the seat beside him. We wound our way down the twisted rocky road among broken towers of stone, through pine and madroña wood. Lichen was spattered like colorful paint on boulders the size of houses. The road was but a spider track that dropped away to the river. A band of white condensation hung above the water, spectral in the canyon air. Dark cliff walls. White water. Blue stone. The green infinity of woods. The red, red earth. And in it all, a couple in native dress, bright as macaw feathers, pushing a child in a wheelbarrow amid alpine spires incised by riverwash. Only the Tarahumara drums seemed large enough to fill this rift.

"It looks like the place where life began," Elena said.

The Tarahumara people, famous for their footraces, can cover hundreds of miles over many days through rocky mountain terrain. They do so wearing huaraches, sandals made from cut-up tires. They live in caves and log cabins throughout a territory made famous by its natural deposits of gold that were forced close to the surface by ages of volcanic activity. This was the setting for B. Traven's novel *The Treasure of the Sierra Madre*.

That night we stayed in an old hacienda lit by candles and kerosene lamps. Roosters cried and musicians played on the veranda. The one with a silver incisor had a pearl-handled nine-millimeter automatic pistol stuck into his belt. The next morning, Carl led us to a flower-filled courtyard and introduced me to our guide, Don Juan, who was part Tarahumara. Even with Carl's extensive experience, he needed a local guide.

Don Juan was sixty-nine years old. He was quite tall by local standards, 5 feet 10 inches in his huaraches. He wore pressed blue jeans and a white button-down Sanford men's dress shirt with a pattern of faint blue and red stripes. His hair was gray and neatly trimmed. His skin was dark and creased with delicate lines. He had fine features and a sharp, mischievous look in his eye. One of the locals told us he had sixty children, but when I questioned him on this point, he brought forth only a smile.

Before the sun had topped the ridge, Don Juan led us up the old Camino Real, past red adobe houses surrounded by blossoming jimsonweed and yellow acacia, high over the wide and rocky reach of the Batopilas River, where horses crossed and drank in midstream amid boulders scattered like dice by the giants of time. The trail climbed high over the river, and Don Juan stopped frequently to show us whatever he found interesting in the natural world: a black fruit that will paralyze you, another that makes your teeth explode. At a curve in the road, he held up his hand, signaling us to stop. We listened to a rustling in the bushes and saw a flash of color. Then Don Juan shouted, "Hey!" and a Tarahumara Indian man emerged from the bushes, a tiny form in skirt and sandals, wearing a battered straw hat and a dirty striped cowboy shirt with pearl buttons. He carried a bow and wore a quiver of arrows. Don Juan and the man, whose name was Juan Ramirez, spoke in the Tarahumara language. Then Don Juan explained that this man was *pajarando*, hunting birds. From the look of the bow, which was not much more than a stick, I was amazed that he could hit anything with it, let alone a skittish bird.

He and Don Juan talked some more, and at last the small man handed Don Juan his bow and two arrows, and Don Juan gave him a few pesos. Juan Ramirez shook our hands in the traditional Tarahumara fashion, palms brushing lightly past each other. (To take hold of a man's hand, as we do in the north, is a sign of aggression to the Tarahumara.) We watched as Juan Ramirez scrambled down the side of the cliff and vanished through the bushes, and then we pressed on.

Don Juan had no profession or trade as such, nor even what you might call a job. For my purposes of research, I thought he was perfectly poised between a very ancient human state and the modern world. He owned a house and a small plot of land on which he cultivated much of the food he and his wife ate. He hunted the wild pigs that roamed the area, and his wife gathered wild nuts and berries and roots. With the exception of a few modern conveniences, such as a fixed home and a cast-iron, wood-burning stove, they still lived much the way hunter-gatherers would have lived about 11,000 years ago, when they began supplementing their food supply with agriculture.

Don Juan was also a kind of entrepreneur. He would find things to sell or trade. He would encourage the Tarahumara to bring him the baskets and blankets and pottery they made, and everyone would profit. To live this way, Don Juan had to know his world. He had to stay close to the earth and be prepared to take advantage of every opportunity it offered. He did not do it through what we think of as science, though his was a rigorous discipline, exacting its toll for failure. He did it as the old ones had, through flexible mental models and behavioral scripts and an accurate sense of himself and his group. And yet, in important ways, he was forced to live as a scientist. He was required to find out the rules of his world and to prove them. His proof was his continued existence. And just as scientific theories can be falsified, so would his theories about his world be overturned if he happened to be wrong. Then nature would say to him, I refute you thus. And would take his life.

We came to a bend in the trail, where a chasm opened up to our left, dropping severely to the rocky river below. Don Juan stopped, fitted the bow with one of the arrows, and let it fly across the open space, laughing with the pleasure of shooting it. I assumed that it was just a frivolous gesture—who cares if we lose an arrow? It cost only a few pesos. But I did not yet know what I was dealing with. I lost sight of the arrow on its flight through the bright sky. But twenty minutes later, when we rounded the path on the other side of the chasm, there it was, lying unharmed in the middle of our trail. Don Juan picked it up with a confident smile. It was a remarkable shot.

He showed us the construction of the arrow. The shaft was cane, the parts bound together with a long fine nerve fiber from the loin of a deer. The fletching was made of buzzard feathers. The arrow told a story of our deep and ancient ancestry. We evolved as creatures who ceaselessly tinker with all the materials of our world to craft improbable and useful devices, things that could not exist without a brain big enough to imagine them. This tinkering with our newly evolved hands helped our brains to grow large in ancient times, as we saw earlier. Discoveries in South Africa show evidence that we were doing this as much as 80,000 years ago; an ochre plaque and snail shells pierced for necklaces bear witness to our symbolic obsession. To say a man makes an arrow seems like no great puzzle. But to say that nature makes a man who makes an arrow is a real mystery. How could nature give rise to brain and hands and then bring them together with these materials to make such a marvel of organization as this arrow?

About 50,000 years ago, perhaps more, a sudden and dramatic change took place in human culture. The complexity and diversity of tools and artifacts exploded. Art appeared. Elaborate rituals attended the burial of the dead, suggesting a new way of thinking. That way of thinking represented a sharp change from dealing with the world in which we found ourselves to rejecting it in favor of one that we created for ourselves. It was a change from living in reality to living in a world of symbols. And with that change, people became, for all intents and purposes, modern humans.

As Ian Tattersall, the head anthropologist at the American Museum of Natural History, pointed out, this wasn't the result of a slow, incremental process from generation to generation. Before that, people made tools, but they were to my mind like the Laetoli woman looking for Safetyland—groping in the dark for something that they knew was out there but that they could not quite comprehend. I see that sort of process in the way the bonobos examine and play with unfamiliar objects and sometimes even find clever uses for them. I can almost hear them thinking, I know this *means* something . . . But they lack the symbols that would close the gap.

For millions of years, people have been picking things up and turning them over in their hands and thinking up sequences of actions that would transform them into something new and useful and eventually even beautiful. There came a moment when a person somewhere in history imagined something that did not yet exist. That person was able to reduce the unmanageably complex world to manageable symbols or mental models and then imagine a sequence of actions that could put the models together to make a new object come into existence. When that happened, it marked a deep change in our history. By definition, that first act of invention completely changed our relationship to our world and our way of comprehending it. For the first time, we realized that we did not have to accept the world as it was. We could live in a world of mental symbols and fashion our own world in their image and likeness. With that outlook—at once fierce and delicate, like the hand itself—the modern gadget-strewn culture that we see around us was launched, and we were on our way to Safetyland. The technique for making the arrow that Juan Ramirez sold us had been passed from hand to hand, brain to brain, for at least tens of thousands of years and possibly much longer. It embodied one of the most profound differences between us and Lucy, the ability to live in a world of symbols.

We came to another bend in the trail, and Don Juan shot an arrow across another drainage in the canyon. Again, I lost sight of it in flight. But we caught up to it in half an hour of hiking, and Don

Juan smiled with satisfaction. Soon we were climbing high up a cleft in the mountain, and the river diminished below. We caught sight of a boy hunting with a bow, and Don Juan called to him in Tarahumara. We left the path and began scrambling diagonally across the face of a steep and jagged slope of iron-red scree. When we caught up with him at last, we saw that the boy, Pedro by name, wore a dark and greasy cowboy shirt over his *tapa rabo*, the Tarahumara kilt. Pedro's father was the same Juan Ramirez who had sold Don Juan the bow, and they looked alike. Pedro's mother had died. "The people die young here," Carl said. "Pedro is an adult." He appeared to be no older than twelve or thirteen. He and Don Juan spoke for a few minutes, and then Pedro led us to the cave where he lived, which commanded a sweeping view of heaven and earth.

The cave was a natural hollow in the cliff face, perhaps 10 by 20 feet, blackened by fire. The ceiling was low. I could barely stand upright. There were bows in various stages of construction and a rough wooden platform on which to sleep. I saw no evidence of pad or blanket. Rain would blow in mercilessly, snow at times. A baby would be vulnerable here. The area was home to vampire bats that came in the night. Don Juan had pointed out the blood along the trail from their feeding. This cave represented a strategy nearly as desperate as Lucy's must have been. This, then, gave hints of how it was 50,000 years ago, maybe more.

Pedro showed me his fishing line, a hook on a bit of monofilament wrapped around a cast-off plastic juice bottle. We stood for a moment, looking out over the vast basin. Don Juan spoke of the "gentiles," who lived two valleys over. "Gentile" was the name the Tarahumara gave to those people who still lived in the old ways and had never been touched by a modern culture. Don Juan said we could go searching, but we'd never find them. They were too shy. I wasn't sure if the gentiles were real or part of a local legend. But when I'm in Mexico, nothing surprises me.

Again, Don Juan drew and fired his bow. I was still unable to track the arrow in flight. We began our long downhill trek. Within ten minutes, I noticed that Pedro was already fishing from the bank of the river far below. When Carlos Castaneda wrote of sorcerers flying in this country, he may have been seeing something like that; for it would have been easy to believe that Pedro could fly, so quickly did he descend the cliff. The explanation of his skill seems obvious: if a child doesn't fall and die in the first few years of life, he will develop mental models and behavioral scripts that fit these steep and rocky slopes perfectly. We can call it whatever we like: procedural learning, simultaneous processing. But the result is the same. Pedro was the Nijinsky of rocks.

Don Juan led us unerringly along pathways that wound up and down hills, until we drew up short around the crest of a small rise topped by numerous low cactus plants. He scrambled up into the cactus, rummaged for a moment as if he'd lost something, and then emerged triumphant with the arrow he'd shot across the valley.

"Did he find the arrow?" Carl asked. "Or did the arrow find him?"

The first shot, I thought: chance. The second, I thought: luck. The third, I thought: skill. And the fourth, I thought: magic. That's how easy it is to come to believe in magic in a place where boys can fly. But it was simply a demonstration of what is possible when you combine mind and body in the process of learning, especially if you start young. It shows the power of models and scripts that are well suited to the environment.

We hiked out across the gentle slope below the mountain, and soon we came to an adobe ranch house flanked by stacks of fireplace logs cut from brazilwood trees. Two older women in print cotton dresses emerged from a doorway beneath the *portal* and greeted Don Juan by name. Under the scorching midday sun, towering, red rosebushes grew beside the porch, fanning up and over the roof. Beyond the yard, which was scattered with rusted machine parts and truck tires that had been mutilated to make soles for sandals, there was a walled garden; and beyond that, the sky-eating mountains rose, mute and tremendous.

We followed Don Juan under the hewn plank awning and passed through a heavy wooden doorway into the dark chamber from which the women had come. There we found a small man behind the counter of a makeshift store that sold white bread and instant masa. Even all the way down here, the old ways were vanishing. We bought Cokes and took the bottles, flecked with ice, out under the awning and sat in wicker chairs. One of the ranch hands, Mañuel, showed us the hide of a bobcat he'd recently killed.

We watched the intense midday light that made the air shimmer above the yard. Great black hornets hummed in the banana trees. A mattress leaned against one wall of the house. Sacks of corn and potatoes were stacked in a corner of the porch, along with skids of masa in two-pound bags for the *tienda*. A plow stood rusting in the yard. A rake and hoe and shovel leaned against it. An axe lay atop the woodpile. We were in the midst of a fabulous material culture, the food and wealth that the evolution of hand and brain had brought us, the symbolic world of our own making. Yet we scarcely noticed it. The human brain is exquisitely tuned to detect novelty. But everything quickly becomes routine, and we stop paying attention.

Mañuel was discussing the politics of prohibition with Carl. A deputy sheriff, Mañuel had recently quit his job because his best friend was killed and his own life threatened. Mañuel was in his fifties, a real Mexican cowboy, soft-spoken, polite, and intelligent. He wore a mustache and a big knife. In his opinion, the danger lay in the ranch weddings, where there were often as many as 200 guests, all of them armed. The weapons ranged from knives and pistols to AK–47s and even hand grenades. Carl leaned over to me and said, "There's always a lot of booze and a lot of grudges."

For the evolution of tool use among apes had yet another effect on neocortical evolution: tools made our culture more dangerous. Chimpanzees and bonobos hit each other routinely in the course of disagreements. The apes I've observed hit very hard, and they mean it. They're certainly capable of killing. The concerted and successful efforts of chimpanzees to exterminate a neighboring group over an extended period of time was documented by Jane Goodall. But introduce tools and accurate throwing, and it's much easier to kill valuable friends and relations. Introduce ulnar opposition and the ability to wield a club or a pointed instrument powerfully and accurately, and it gets even worse. Clans that killed their own members would have been more likely to die out. A given clan would have been more likely to survive if it developed self-control and better ways of resolving disputes. Though we can't know how they died out, there have been many dead-end species in our ancestry, such as *Paranthropus boisei* and the Neanderthals. And if they didn't kill off their own kind, people like us may have simply wiped them out. As Jared Diamond wrote in *Guns, Germs, and Steel*, killing strangers was a way of life for traditional people until very recently. He describes how when strangers in New Guinea meet, they engage in "long discussion of their relatives, in an attempt to establish some relationship and hence some reason why the two should not attempt to kill each other." Only in the last 7,500 years, Diamond says, have people learned "for the first time in history, how to encounter strangers regularly without attempting to kill them."

The need for self-control may have been enough to encourage the establishment of rituals that slowed the pace of many potentially volatile activities such as eating. Humans may have favored wearing clothes, even in warm climates, and mating in private, because it provided a buffer for strong emotions. Fighting may have gradually turned to less and less lethal forms of sport over time. Those animals with more power to think and reflect on their own actions would have survived in greater numbers, in part by not killing their friends and relations wholesale with their clever skills and crafty tools.

While it's easy for an emotional response to overwhelm the frontal cortex and rational thought, the restraints of the frontal lobes can be surprisingly effective as well.

Our newfound restraint forced the frontal cortex to grow larger still. More restraint, better planning, and increased complexity in how to manipulate objects and images all led to something akin to what we are now: thoughtful, communicative, and capable of mentally planning out complex sequential tasks, sentences, and ideas. Without this ancient compulsion to follow a symbolic complex of programmed actions, Juan Ramirez would have found it impossible to assemble—or even to conceive of assembling—the arrow shaft with nerve fiber from a deer and feathers from a buzzard. Don Juan's ability to fire that arrow with such accuracy was a matter of learning through sequential processing, as a child, how to transform his actions into a reliable behavioral script.

Infants are born with no dominant hemisphere. They use simultaneous processing exclusively for such tasks as reading their mothers' emotional states. Only when they begin speaking does their gray matter explode in the left hemisphere and begin the life of sequential processing, the job of logical, stepwise thinking. Then the struggle for control of behavior begins. Or, as Steven Pinker put it, "The left hemisphere constantly weaves a coherent but false account of the behavior chosen without its knowledge by the right." And between the two we fashion and live in a world of symbols, our mental models.

We live in the balance. But big trouble comes when the frontal cortex, with its knack for rational thinking and restraint, is disabled. And in the ranch weddings of the Copper Canyon region, the big culprit, as Mañuel was explaining to us, was the mind-bending drug alcohol. The area was dry. Liquor was illegal. Making moonshine was a gold mine because of that. Prohibition, Mañuel opined, was partly responsible for the proliferation of "goat's horns," which is what the locals call the AK-47. The trouble with alcohol is that it happens to be an extremely effective drug for taking the frontal cortex off-line so that the emotional brain can run free. Alcohol thus transforms people into their very ancient ancestors. Unfortunately, alcohol doesn't disable the hand and arm as quickly as it does the frontal lobes. In addition, our ancient ancestors had far less lethal tools.

We pressed on up the trail, bushwhacking part of the way. Don Juan flowed effortlessly through the vegetation, and I tried to follow where I saw him go. But every thorny branch seemed to catch my clothing and hold me back. This was the difference between us: he was completely adapted to this place. He had formed his behavioral scripts in his youth, and passing through this terrain was as easy for him as navigating city traffic was for me. He wouldn't have dreamed of trying to go where I lived. On the other hand, being a product of Safetyland, I thought it was my right to go blithely wherever I pleased.

We came to a trail 20 inches wide that traced the edge of a buttressed cliff. We followed it until another trail intersected it, and there we met a man. He was a thin and wispy Mayo with a thin and wispy mustache, a few teeth like brown stones held loosely in his mouth. This faint shade of a man, Miguel, smiled and greeted us happily and touched hands with us. Miguel and Don Juan fell into an animated conversation about the beehives that hung on the cliff straight above us. The nests were oblong, perhaps a foot in length, and looked as if they were made of gray paper. Don Juan said that the bees had no stingers, and Miguel said they made excellent honey. Don Juan's eyes glittered as an idea formed. He told Miguel to bring the honey, comb and all, to Batopilas, and he would buy it from him.

Farther along the trail, we found another Tarahumara cave dwelling. It was nothing more than an overhanging part of the cliff, offering scant protection from the elements. The ceiling was blackened from countless fires, and the floor was littered with stones the size of loaves of bread. A cache of firewood had been piled in the back. Someone had kept a cow in there at one point, and every surface was spattered with liquid dung, which had dried like plaster over rock and wood alike.

We sat on a ledge and opened our lunch, which one of the women at the hacienda had wrapped in linen early that morning. In the great converging distances, a hawk turned slowly, giving substance to the air. Carl, Don Juan, and I each gave Miguel a taco or a biscuit or an apple, so that he could eat with us. He invited us to come and watch him make moonshine over in the next valley. Don Juan said that Miguel was one of the best moonshiners in the region. He lived, he said, "about an hour north." Carl turned to me in an aside and said, "That would be more like three hours for us."

We finished our lunch, drinking water from the bottles we carried, and Miguel and Don Juan talked of water—where it was, how good it was, how it was found, how reliable, and how to reach those special springs that never dry up. Vast aquifers lay beneath this dry land, and knowing where they reached the surface meant the difference between life and death. Knowledge was being exchanged, secret and imperative, intended only for the tribe. The words they spoke gave life and thus were sacred. There are places, Don Juan whispered, where water comes out of the mountain as cold as snowmelt. That was the best water.

Their talk and their keenest interests always focused on what was needed for survival and on new ways of making and remaking their world through technology, finding and extracting resources from it through constant attention and analysis. Their joy, their lives, were all wrapped up in this passionate pastime. In ancient times, this would have been true of the men more than the women, who, as scientists like Robin Dunbar believe, would have been building and refining the essential social fabric by talking about people. The earliest evidence of people using fire suggests that men made hot fires that burned fast. They used them and then presumably went on their way. Women made cooler fires that burned all day, suggesting that they spent more time around them. Even though burdened with caring for the children and the home fires, the women would hunt when the men were away or were ineffective.

In the world that Don Juan and Miguel inhabited, like the world of traditional people, at every turn in the path there were dozens of things that could kill you, dozens of things that could feed you, dozens of things that could delight or entertain you. They would have had a hard time understanding certain concepts we bring from our world, such as acquiring something in order to throw it away.

The rich diversity that Don Juan and Miguel experienced in their surroundings has been replaced in our culture by endless duplication. The advent of our technical culture, born in the Industrial Revolution and supercharged by postwar entrepreneurs such as those from the class of 1949 at the Harvard Business School, has brought us the ability to blanket the world with millions of identical products, from plastic cigarette lighters to computers and cars. The very fact of being able to produce so much, however, means that people have to be induced to buy more and more, which in turn means that we must discard what we've acquired. It was described well by the influential paleontologist and popular author Stephen Jay Gould, when he referred to "something precious, something fragile, and something sadly lost when institutions become so large that the generic blandness of commercial immensity chokes off both spontaneity and originality." This state of affairs came into being in my own lifetime. This culture of trash has a lot to do with how we have come to drop our guard. The mental models we form blind us to its effects.

We are evolved to relish novelty and then to figure out a use for the new thing we discover. We are equally well evolved to get tired of the new thing and want more. We must continually work harder and have more to achieve the same level of satisfaction. All apes are like that to a degree. The difference is that we're the only ones clever enough to make the dream of plenty come true.

Miguel smoked after lunch and told us that during the Mexican Revolution Pancho Villa's men had dug trenches near this place during battles, and that some of them had had to climb to the top of the butte just above us in order to defend themselves from the *federales*. Don Juan said he had known one of those men when he was a boy. The old warrior, who had lived in Batopilas, was on top of the mesa one evening with Villa's troops when the *federales* attacked from below. Villa's men had been kneading masa into balls to make tortillas and were so terrified during the attack that, without thinking, they threw their masa balls at the *federales* to defend themselves. Don Juan and Miguel couldn't help laughing at the grand history of Mexico unfolding in a food fight. The old revolutionary had told Don Juan that he had shit himself from fear as he ran down the mountain to safety, and we all agreed that in battle, it was no shame to throw masa balls, nor even to shit oneself.

Throwing became an essential behavioral script for hominids a very long time ago. Under stress, we revert to what is primary, and often the horrors of war can be understood that way. War does not ennoble men. On the contrary, under the prolonged stress of combat, men revert to behavior that we might expect from a much earlier hominid: rape and the mass slaughter of the innocent. War pushes us back in evolutionary time and makes us beasts. It forces us to use our most ancient and outmoded models and scripts and act out the worst elements of our groupness.

The imperial Japanese army held killing contests in Nanking, China, in 1939, to see who could behead 100 people in a single day. In an astonishing demonstration of how groupness can become the servant of evil, a popular Japanese newspaper ran photos of the killings and the headline, "Contest to Kill First 100 Chinese with Sword Extended When Both Fighters Exceed Mark—Mukai Scores 106 and Noda 105."

If more recent and intimate evidence were needed for our shared proclivities, the fact that we are killing our own children by sending them to war at this glorious stage of our evolution should provide it unequivocally. My father went to college watching his friends go off to Europe to die. I went to college watching my friends go to Vietnam to die. My daughter Amelia went to college watching her friends go to Iraq to die. My grandfather didn't go to college, but he wound up in the U.S. because he was running from the Mexican Revolution, which killed one out of every eight Mexicans in the country before it was over. He was just about Pedro's age. No human generation has yet known peace. The burning question for our survival as a species is: Can we leave the ape behind and grow up?

After Miguel had smoked, he walked with us for a time, but soon the trail divided again. There the moonshiner brushed our hands, said good-bye, and took off up the other side of the mountain at a good clip. We went on around the rock outcropping that formed the mesa which had towered above us the whole day. Soon we found ourselves climbing over loose rock again, even more precariously situated than before, as we made the final pitch to the top. The rocky trail, which the locals called The Snails, narrowed as it angled upward, and finally shrank to a width of about a foot and a half, staircase steep, with the butte's flank leaping upward to my right and a drop of 600 feet to my left. I spotted Miguel at the very top of a nearby peak, squatting above a precipice, watching our progress with a wan smile. He waved to me, an amusing little man who had somehow leapt 1,000 vertical feet in minutes. As I struggled upward, I saw him rise and lope away into the piñon scrub, bent over, his hands swinging low at his sides.

When at last I reached the top, panting and sweating, I found myself in a grove of ancient stunted oaks in a well-organized stone ruin. All around us, obscured by undergrowth, were the remains of an ancient city. The low stone outlines of houses were unmistakable. They were perhaps 10 feet square, some joined in rows to make a city street. Some were round, perhaps storage bins or granaries. There was a clearly delineated plaza or town square, flanked by verandas or dance patios. It was no accident, no temporary settlement, but a well-wrought city set on top of a mesa in this nearly inaccessible location.

Carl became agitated, darting through the oaks, exclaiming, "No archeology has ever been done here. This is really a find. This is amazing." For a long time, we examined the ruin, touching the stones that held so much history. Don Juan called it the City of the Giants. The impulse to call those people giants was not unique. When the Germanic peoples of the Dark Ages, caught between the fall of the Roman Empire and the Norman Conquest, came upon the ruined city of Bath, they called it "the work of the giants." If you've forgotten your skills and crafts, the assumption made sense. The stones were too large to lift.

History shows that when we forget our skills and crafts, we go back to an earlier time and start over. We grasp with words what we have forgotten with the hand, and then we pull ourselves up once more. I imagine a time in our not-so-distant future when our descendants will come upon the ruin of St. Louis or Detroit, and one of them—illiterate once again but with the gift of the gab—will improvise a poem about the kingly thing that was, that shattered city, the wallstone once bravely bound in iron, and call it the Work of the Giants.

Chuy drove me and Carl and Elena out of Batopilas and up to a lodge in Cusarare, high on the rim of Copper Canyon. I was on my way home. The drive took all day, and when I awoke the next morning, the fire had gone cold in the small log cabin where I stayed. I stumbled shivering into the dawn. Elena was still asleep. The irrigation ditches were skimmed with ice that glinted in fragile trapezoids like the iridescent wings of some prehistoric bird lying half-buried in the sand. Pennants of smoke stood windless and ghostly atop the huts of surrounding ranches, and a piney odor reached me on that thin, still air. In a neighboring corral, a man in a burlap apron was bent over in first light, honing a great knife against a stone. A startled group of pigs milled around him, pig-muttering half-aloud in what sounded like the musings of a lunatic. A small boy emerged from the dark door of the shack and stood for a moment, sleepy in the dawn, his breath ghosting on the still air in imitation of the smoke from the woodstove. He stepped barefoot into the muddy yard. A blond and woolly pony trotted over and sniffed him lovingly. The boy joined the man in the burlap apron, and the pony followed.

Carl came out of his cabin, and together we crossed an immense and gently sloping dome of lava tuff, then climbed at a steep angle through a forest. Here at 8,000 feet, the long pine needles were rimed with frost, and the first light made the tree trunks seem like dark and spectral shapes that crept through a world of scintillating glass.

Half an hour later, we topped the mesa and crossed through more pine forest to the other side. I was wary of allowing Carl to lead me over a cliff, but the volcanic rock offered easy handholds. We descended the cliff to a path that led between boulders to a complex of cave dwellings. It was a small city built into the cliff. There was even a stone corral for animals. The cave city formed a sheltered veranda that overlooked the entire valley. I'd seen ones just like it above a river in central France.

We reached the entrance to the first room, and Carl showed me how the wooden door worked, with its hidden locking mechanism, a wooden bolt that slid out. Wooden pins formed hinges for the small hatch, through which we crawled into a chamber. We had to wait for our eyes to adjust to the dark before we could move around. There, on a crude board, we saw a perfect ancient pot the size of a grapefruit. It had been painted with delicate designs, and Carl reported that nothing had been touched since his last visit to this room some months before. Looking at the fragile vessel, Carl and I shared the same silent thought. At length, he said, "Someone's going to come up here sooner or later and find this pot and break it. It always happens. It's kids who find this stuff, and to them it's just something to be thrown against a wall."

For the longest time we struggled with the impulse to take the tiny pot and thereby save it from destruction. Desire struggled with restraint. I wanted to possess it, to believe that it had been put there for me. But it had not. It was just there, part of the world, and I had to bear witness and let go. At last we closed the door and went on. To smash the pot or to take it as loot—that was the way back. To visit our past, to walk in that ancient world without touching it, that was the way forward. We did not want to step back into the world of loot, for that ceramic vessel was not loot. It was the concentrated essence of human history, condensed into a single subtle spheroid object, caught in time and light. It was an idea made manifest, an object, both practical and symbolic, that did not exist until imagined in a human mind, a mind that then found the materials and the sequential steps that brought it into being with the remarkable skill of the human hand. I was fortunate to have seen it. What I took away from there was far more valuable than the pot: memory, and the dignity that only a human can have when he rises above the ape.

After our trip, Elena wrote about her travels. "Experience makes the person," she wrote. "In certain people I have seen something that makes me realize how much there is yet to know. I once met a photographer named Candace who had seen the whole world. Just looking at her, I could see that she understood something essential. Most people let their world pass them by. There is a difference between the wanderings of Odysseus, who saw what he saw because he was lost, and Candace, who went into the world because she felt a need to know everything. Some people have a tremendous curiosity, a need to know things that are hidden." It is in the very impulse that Elena describes that the real hope for humanity lies.

Eight

Pleonexia

The long struggle that the Laetoli woman and Lucy were engaged in, the bitter legacy that was passed on to every generation, had all but ended for the American middle class by the time I was old enough to question the largesse of my world. What a remarkable thing to bear witness to, like the discovery of how to make fire. Millions of years of struggle, and here I am, safe, warm, and well fed at last. How did it happen? What's the catch?

The year before my birth, President Harry Truman signed into law the Employment Act of 1946. It promised, among other things, to promote "maximum employment, production, and purchasing power," as well as "the propensity to consume." Mine was the first generation in the history of humanity that did not have to deal with scarcity of at least some resources at least some of the time. A state of plenty had been enjoyed by various elite groups throughout history, but never had it come within easy reach of so many. And none of them had had access to the medical care, the transportation, the protection, the weaponry, or the ability to communicate that now gives us the dangerous illusion that the world is truly our own.

My father's father, a successful small-time retailer, lost everything in the crash of 1929. He understood that failure was, indeed, an option. People of his generation believed in their own mortality. They did not easily fall prey to a vacation state of mind. My parents knew hunger as children during the Depression—if not their own, then that of their neighbors. But by the time I was ten years old, things had changed dramatically. My father was a lowly postdoc assistant professor in the Texas backwater town that was Houston. He had a growing family (I would have six brothers by 1960) and a mortgage. We were very far from rich, but I wore the starched uniform of a private school. My concerns were not whether I'd be fed or clothed or sheltered. By the time I was thirteen, my friends had a stable of cars that I could drive, and gasoline was 31 cents a gallon ($2.22 in 2008 dollars). My concern was being cool.

In the early days of my childhood, just as the fuse of consumer capitalism was being lit, I lived with my loud and playful family in a cold-water flat on the north end of St. Louis. When the wind was right, we could hear the Cardinal fans cheering from Sportsman's Park. We were just barely entering the middle class then, yet underneath the Christmas tree each year there were presents for all. We had not just plenty of what was necessary; we could expend our resources on what was frivolous.

My mother's younger brother Bob worked for Sears, Roebuck and Company and always had the latest gadgets—the washing machine, the lawn mower, the electric mixer, the toaster. One weekend, he and my father installed hot water heaters in the duplex our families shared. I was too little to care, but what a luxury it must have been for my mother to turn on the tap for a hot bath in winter instead of having to boil water. All of these advances were greeted with enthusiasm. Each one made life a bit easier or more delightful. With each new invention, we expended a little less effort and burned a little more coal.

On one particular evening during this time, I had been playing in front of the two-flat and went inside to find that no one was there.

We all lived on the same block, or nearly so: our family and my mother's parents, my aunts and uncles and cousins . . . So, finding no one home, I simply went upstairs to Uncle Bob's apartment. All was silent. It seemed that no one was home at his house, either.

Wondering what could have become of them, I crept about the flat, sniffing, searching, listening. Then I heard voices. They weren't ordinary voices. They were beset by static, as if they came from a badly tuned radio. At last I made my way to the parlor. There I found everyone seated with their backs to me as I entered. They were in big easy chairs and squatting on ottomans and sitting rigidly on folding chairs that had been dragged in. They were all staring at a piece of art deco furniture with a bright, circular, bathyscaphe screen set into it.* On the glowing screen were images of very small people and horses and cars. A world in miniature. My family members, once so loud and joyous, now sat in the dark, utterly silent, transfixed by this device. Something had taken over their attention. Even though I was only three or four at the time, I knew that I was seeing something momentous. I knew it deeply, in the way that children know things before they are permanently distracted from paying attention.

My life has straddled two cultures, because my mother is from German and Irish stock in St. Louis, while my father's relatives are Texas Mexicans and Native Americans. Back in San Antonio, I saw my relatives go through a similar transformation. I knew nothing of privation, so to me it was curious to see my great-grandmother, Ramona Hernández, grinding corn on a stone metate and raising chickens in her backyard. I didn't understand that her way was the only way anyone had survived when she was my age. I remember how frugal my grandmother Rosa, her daughter, was. I thought it quaint that she ate the cartilage and even parts of the bone when we had chicken for dinner. Their indelible model of the world dictated

*I'd later learn that it was a 1950 Zenith G2438R with a black-and-white screen 16 inches in diameter.

that one could not afford to throw things away. The deep physical lesson learned from hunger was clear: if you didn't want it all, you wouldn't get any of it.

How quickly we forgot. After the end of the war, it didn't take long. When my father was a boy, each child in his neighborhood could expect to get one bicycle. That was for life. I am trying to think of something—anything—that I have for life today. Not a suit of clothes, not a car, not a house, not even a city. My grandparents lived in the same neighborhood all their lives. And my grandfather's grandfather was born and died in the same Mexican village without ever venturing outside of it.

As I grew up, more and more people in my father's family moved out of the old barrio with its dirt streets and chickens and took up residence in the suburban housing complexes. They bought newly built, air-conditioned homes with washers and dryers. The stores sold instant masa for making tortillas. There was no more grinding corn. My great-grandmother died. Her metate lay cracked and weathered in the backyard garden. My generation grew up not even knowing how to use it. Then everyone stopped making tortillas altogether, because a factory had opened that mass-produced them. We lost our crafts and skills. There were no chickens or donkeys, and pretty soon we didn't even have to get up to change the channels. We dropped our guard. And it was the most natural thing to do.

By the 1960s and probably much earlier, the question had already become: What is the way back? We can't give up the plenty, the convenience of our lives, the technological advances, our protection from predators. I don't want to raise chickens. We can't suddenly become Lucy again—and we wouldn't want to. Few of us would last a week trying on her way of life. Would I want even my great-grandmother's much milder form of privation? No, not even that. We can't go back, we must go forward.

But what is the way forward? I know what it isn't. It's not, as we once believed, plenty to eat and a home with all the modern conveniences. It's not a 2,000-mile-long wall to keep Mexicans out or more accurate weapons to kill them. It's not a better low-fat meal or a faster computer speed. It's not a deodorant, a car, a soft drink, a skin cream. The way forward is found on a path through the wilderness of the head and heart—reason and emotion. Thinking, knowing, understanding. And for me, at any rate, it begins by trying to comprehend the world and myself in it. It's the most important difference between us and Lucy: that we can know and think in new ways. Real knowledge and understanding can be ours. Think of it as American Zen. As with everything else in our culture, all we have to do is want it.

The American philosopher Eric Hoffer wrote, "Faith, enthusiasm, and passionate intensity in general are substitutes for the self-confidence born of experience and the possession of skill. Where there is the necessary skill to move mountains there is no need for the faith that moves mountains." When we have everything done for us by others or by machines, we lose our skills, our crafts, and our need to understand our own world. This cuts us off from the opportunity for personal action leading to self-confidence and self-esteem. Thus disenfranchised, we are apt to make mischief for ourselves and others. Hoffer's way of life contains some important lessons for us all.

Hoffer worked as a migrant farm hand in California for more than a decade, and then as a longshoreman on the docks in San Francisco until his retirement at age sixty-five. But he had a hunger for learning. As a migrant farm worker, he had library cards in small towns up and down the coast so that he could read whenever he wasn't picking peas or cotton or tomatoes. He was literally crazy about learning. He had a Santa Fe Institute kind of mind, teaching himself chemistry, physics, mineralogy, mathematics, geography, and history. One day at SFI, I mentioned to Cormac McCarthy that I was rereading Hoffer, and he said, "Oh, he's my hero."

One day, working in a nursery transplanting tomato seedlings,

Hoffer realized that he didn't know why plants grow with the roots down and the leaves up. He saw the fact that he'd neglected botany as a dire emergency. Without delay, he went to the office, collected his pay, quit his job, and hopped a freight train to San Jose, where there was a decent library. There he read Strassburger's gigantic textbook on botany.

Several times during his life, people recognized Hoffer's genius and tried to recruit him into universities, but he kept his distance. In this unusual way, he gained a genuine understanding of his world and of human nature. He became what he called "a natural thinker." I view that as much more than just a worthy goal. It is an essential pursuit if we are to move forward.

Lucy and the Laetoli woman were trying so hard to comprehend their world. Since then we have gained astonishing brainpower, which we have used to erase the most towering challenges of Lucy's life: we have plentiful resources and we have vanquished our predators. But our bounty comes with unintended side effects that we now face. They are difficult to face, because our stubborn mental models stand in the way. I want to show you, on a very small scale, a side effect of our efforts and how our models and scripts can trip us up. This is one of the costs of being carefree, as well as the price of giving up self-reliance and seeking the perfect protection of others. This is one of the taxes we pay on our triumph over Lucy's predicament.

Airbags were introduced in automobiles in the early 1990s, not to make driving safer but to make crashing safer. That's an important distinction, because instead of starting from the assumption that we could do better as drivers, it was decided that we could not. We were not teachable. We would simply keep crashing no matter what. So we had to be protected—passively, as it were—during those crashes.

An airbag is a small cannon that is aimed at your face when you're seated in a car. It is set to fire when you crash. The expanding gases from the exploding cannon shell inflate a tough plastic bag, which is intended to act as a pillow to stop your forward motion. All well and good. But soon after airbags were introduced, they began killing children who happened to be in the front seat, where a mother would naturally put a baby to have him within easy reach. The airbag was designed to stop a full-sized adult. The babies were small and light. The explosive force of the airbag crushed the babies and broke their necks.

There's the catch: everything comes with a cost. An attempt to improve safety had created a new hazard. Warnings were issued. News stories were written. Cars were manufactured with bright yellow stickers saying, "Warning Airbag!" Laws were passed requiring parents to put children in the backseat. The initial assumption that we were too stupid to learn to become better drivers had now met the ugly facts. In the face of a new and deadly technology, we'd just have to be taught something else. Could we handle it?

Actually, people can be very smart when given the chance. They did put their children in the back. But now that the new technology had been subdued, we had to contend with another deeply human trait: the making of mental models and behavioral scripts. The presence of a baby next to you in the front seat, cooing and squirming, might initiate action based on a script you could call "Taking Care of Baby." But without the baby in plain sight, it would be easy for that same system to unconsciously reach a new conclusion: the baby's not here. When a new baby arrives, it takes a bit of time to create the reliable mental models and behavioral scripts that most parents find familiar. They are manifested in the constant nagging question: Where is the baby? But before that script is created, the absence of the baby from immediate view in the car can initiate an older script that doesn't include the baby as part of our model of the world. None of this requires deliberate thought.

So, acting on a outdated script, parents walked away from their cars, leaving the children inside. The temperature in a car, even with the windows cracked, will rapidly rise to dangerous levels, even in cool weather. The result was that many babies died of hyperthermia—they were slow-cooked to death.

Between 1990 and 1992, when there were few airbags, there were only eleven such deaths. Between 2003 and 2005, when almost all new cars had airbags, 119 children died from being left in cars. The frame of mind we're put into by passive restraints ("passive" means that we aren't responsible), and the convenience of a device to lock the doors of the car remotely, serve to make these kinds of mistakes more likely.

This is but one small cost of Safetyland. When we are not living examined lives, when we aren't paying attention, when we are not practicing self-reliance, other forces slip in to dominate our lives, our behavior, and ultimately our fate and our future. Relying on others and losing our own abilities has made ours a fearful and vindictive society. Societies, like individuals and institutions, build emotional systems. Shocks to the system can accumulate and lead to overreaction. For example, in 1982, someone put cyanide into bottles of Tylenol and replaced the bottles on shelves in stores. Seven people died as a result. The death of seven people is not a very large event in our country, but that event produced a deep change in our cultural emotional system. Every conceivable kind of package was quickly sealed up beyond any normal individual's ability to open it. That's just one shock to which we overreacted. Safety measures work like technology: they suffer from a ratchet effect. They can only go in one direction. Once you invent the car, you can't go back to the horse. Once you seal the bottle ("for your protection"), you can never again sell an unsealed one. Over the years our society was shaken by many such shocks, from the Columbine High School massacre to the collapse of the World Trade Center. Each shock led to more control, and our society began to lose its flexibility, its adaptability.

We no longer know how to react to ordinary events. In White Plains, New York, in July 2007, eight students taped alarm clocks to the walls of their high school as a prank and were charged with the felony of "placing a false bomb," which carried jail time. This is not an isolated incident. It is a widespread trend in this country toward "getting tough." In December 2007, a ten-year-old girl in Ocala, Florida, was sent to school with a lunch that included steak. In her lunch bag was a small knife with which to cut the steak. Her teachers at Sunrise Elementary School saw the knife and called the Marion County sheriff. The girl was arrested and faced felony weapons charges. When a society turns on its own children, it is no longer functioning normally. Getting tough is only necessary when you have already lost control. And the fact that we can no longer distinguish between terrorists or violent criminals and our own children—or for that matter, between eating utensils and weapons—is an ominous sign. By this reasoning, the baseball team would be arrested for wielding clubs. But if you can no longer think, you can't be reasonable.

It is the failure of thought and reason that leads to such outcomes, in which people have followed a seemingly logical path to reach complete nonsense. Hence the phrase "intelligent mistakes." When people abdicate responsibility and come to rely on a system of rigid rules, matters can take an ugly and dangerous turn.

When reporters came to question the authorities at the Sunrise Elementary School, they were told that the teachers had no choice about what they had done to the little girl. "Anytime there's a weapon on campus, yes, we have to report that," said a spokesman for the county school system. A spokesman for the Marion County sheriff's office defended his actions against the girl in the same way: "Once we're notified," he said, "then we have to take some kind of action." The most frightening element of this drama is the refusal of ordinary people to think at all. Their answers are reminiscent of those given by officers of the German Reich during the Nuremberg trials, in which they said, as if it were obvious, that they were following orders. And of course, they were.

Anytime one brings in the Nazis for comparison, it seems like overreaching. Nothing, we think, can compare to the Nazis. They were the ultimate evil, never to be matched again. But there is another point about the Holocaust that is often missed. Throughout this book, I have been discussing how thinking clearly and paying attention in new ways can influence our behavior for the better. I have been arguing for the importance of having a deeply ingrained habit of questioning and analyzing ourselves, our environment, our behavior, and the behavior of others. Nothing in modern history supports this point better than the story of the Holocaust.

From 1941 to 1945 SS Obersturmbannführer Karl Adolf Eichmann was in charge of the Department for Jewish Affairs in the Gestapo. As head of the notorious Section IV B4 for "Jewish affairs and evacuation," Eichmann ran a vast and smoothly functioning system for deporting Jews to death camps and stealing their property in the name of the Reich. His operation, first established in Vienna, was so efficient that it was replicated in Prague, Berlin, and elsewhere. His organizational activities directly resulted in the murder of at least three million people. In 1960 Israeli Mossad agents kidnapped Eichmann in Argentina. He was taken to Jerusalem, tried, and on May 31, 1962, he was hanged.

In 1963, the political philosopher Hannah Arendt published a book on his trial, *Eichmann in Jerusalem: A Report on the Banality of Evil.* The shocking point of Arendt's account was that Eichmann, as she put it, "was not Iago and not Macbeth." He was not a man of evil motives. In fact, "He had no motives at all." After exhaustive research into the matter, Arendt concluded that Eichmann "*never realized what he was doing.*" (The emphasis is hers.) She writes that "such remoteness from reality and such thoughtlessness can wreak more havoc than all the evil instincts taken together." The famed Israeli journalist Amos Elon took this a step further, declaring that "evil comes from a failure to think." He pointed out that one of the great failures of Eichmann's trial was that "his essential brainlessness was never even brought up or discussed. It wasn't discussed partly because it was so hard to grasp."

The inability to think clearly is an unintended side effect of our most glorious achievements. This is not an argument for doing nothing or for achieving less. Neither is it a Luddite rant against technology. I happen to love clever gadgets—my Canon digital camera and airplanes that fly upside down. I embrace modern conveniences, such as the furnace that keeps me warm without my having to stoke it with coal every few hours. Mine is rather an argument for being aware of the costs of these wonderful things—and perhaps for planning how we will offset those costs—as we enjoy our great way of life. It is an argument for being curious enough to pay attention. Because the cost of being unaware may already have grown too high for us to bear.

Since the Industrial Revolution, we've raised the amount of carbon dioxide* in the atmosphere sharply, causing a sudden jump in the temperature of the earth. You don't have to be a scientist to comprehend what's going on. I went up to Glacier National Park and farther up to Alaska and saw for myself what's happening to the glaciers: They're disappearing. The glaciers in the Himalayas are melting at a rate of 30 feet a year. If you fly from Europe to the U.S., you can see the Greenland ice sheet melting. (When it melts completely, the water locked up in that ice will cause the sea to rise more than 20 feet.) This is what's occurring after a mere one-degree rise in temperature (0.6 C). I find it hard to worry about the next century when I'm having such a good time living large in America, but if we keep increasing our use of carbon at the present rate, the earth will heat up considerably more. Put another way, if this trend doesn't change, my son Jonas, who turned six around the publication date of this book, will grow up in a world of increasing disruption of the weather, agriculture, freshwater supplies, and the oceans. He may well witness the largest mass die-off of humans in history. The effects of global warming have been well described elsewhere. Those effects can be expected

*Only about 13 percent of that carbon comes from cars and trucks.

to take back a lot of the advantages we've gained since the Industrial Revolution. My generation could be the last to enjoy the largesse I've seen come into existence in my lifetime. Or we could change.

I don't mean to imply that Lucy or any of our more recent ancestors were any more responsible about their environment than we are about ours. I don't regard any of the people who came before us as righteous and loving stewards of the earth. They wasted resources with profligate inefficiency. There is no noble savage to look back on longingly. On the contrary, I believe that we have the potential to be the noble ones. But even though our ancestors wiped out many species that I wish we could see, and cut down forests that I wish were growing today, they didn't have the means that we have for disruption on a global scale.

Many of our practices start out as good ideas and just get out of hand. Over time, the trend seems to be that the smarter we get, the greater the damage. The native inhabitants of North America taught Europeans to bury a dead fish with the corn seed to fertilize it. That's because the decomposing fish contains nitrogen and phosphorus, which will nourish the corn plant. It sounds like a good idea and very conservative in terms of its impact on the environment. But an even better idea—or so it seemed—was dreamed up by Fritz Haber, the German chemist who in 1908 figured out how to create ammonia artificially. That ammonia is the principal ingredient in the fertilizers that started the Green Revolution and made modern agriculture possible. Just more of a good thing, it seems.

But it, too, had its unintended consequences. The bountiful food supply encouraged a global population explosion, as the overabundance of grains produced by that nitrogen was shipped around the world. The more we ate, the more numerous we became, and the more food we needed to grow (and the more carbon we produced). Fertilizer was the only way to grow enough, but the nitrogen ran off fields and into rivers. It has completely changed the chemistry of life around the world. Fresh water in the United States contains twenty times more nitrogen than it did 150 years ago. The rivers flow into the sea, where plants thrive on nitrogen just as the Indian corn plant did. These so-called algal blooms die, and bacteria eat the decaying plant matter and use up all the oxygen. That leads to the death of larger species, such as fish. Eight thousand square miles of the Gulf of Mexico are effectively turned into a lifeless zone every year through that effect. If one fish is good for producing nitrogen, is the equivalent of a trillion fish better? This is just one example of how a perfectly reasonable impulse, when powered by enough cleverness, can turn into a nightmare.

There are at least 150 such lifeless zones in the seas of the world today, and that number doubles every ten years. A growing number of them, such as the one in the Baltic Sea that covers 27,000 square miles, are permanent now. Dead sea is a very bad thing.

We can view our present situation as one that simply evolved from the natural functioning of the human brain and the very reasonable efforts of people, beginning long ago, to ensure their own survival. Viewed that way, our behavior now is not much different from what it was a million years ago. We squander, just as our ancestors did. Unlike our ancestors, who could merely make a few species go extinct, we are—as Edward O. Wilson believes—in the process of causing a sixth mass extinction of plants and animals on earth.

We can change. More to the point, we must change.

In September 2007, Daniel Schrag, a MacArthur fellow and a professor of geochemistry at Harvard, delivered the Ulam Memorial Lecture Series at the Santa Fe Institute. For three nights, he carefully explained in plain language how the earth handles carbon and how we have systematically (if unconsciously) worked to upset that system. The conclusion was clear: any level of carbon in the atmosphere that is above 280 parts per million is dangerous. That was the level just before the Industrial Revolution. Largely because we burn coal, that level now stands at 384 ppm, and it is certain to rise to at least 500, even if we do everything right. If we do everything right, the earth will continue to grow warmer for decades to come. So there is still a very large tax to be paid on all the coal we've already burned, to say nothing of what we're continuing to burn. The Chinese are right now building coal-fired power plants at a breakneck pace of one per week, and there isn't a clean coal-fired plant in the world yet. Moreover, history shows that humans have never done everything right. So there is a serious question about whether the vacation state of mind that we've created for ourselves (and that China and India are so eager to embrace) has already led to an irreversible process of runaway global warming. In the meantime, we'd better start thinking seriously about how we can undo the damage we've done.

Our allies in that effort can be—indeed, must be—big business and even big oil and the Pentagon, the very entities that many people have traditionally regarded as the main obstacles to improving our stewardship of the earth. The way we are handling our use of energy now is very much a mirror of the way *Homo erectus* handled his resources when he drove herds of elephants to their deaths by setting grass fires, killing the whole herd to harvest a few animals. We're much smarter now, and our technology is so much more advanced that we can use a fraction of the energy, even while we reap a greater benefit from it, whether that benefit is defined as corporate profit or a comfortable way of life or even a larger gross domestic product.

During the oil crisis that spanned the years from 1977 to 1985, the economy grew by 27 percent, even though we used 17 percent less oil. We cut imported oil by half. What changed? The high price of oil—the punishment for our behavior—dictated a change in our mental models and behavioral scripts that told us how our world worked. We received a shock equivalent to the one I received when I tried to pick up a rattlesnake. The trouble is that by conserving energy, we drove the price of oil down again. Then, during the next fifteen years, when we weren't punished for wasting energy, we gradually returned to our old ways. That's because we weren't reflecting on our own behavior. We were not thinking. We were just doing what came naturally.

But those years from 1977 to 1985 proved that we can change and be more efficient, even while living well and growing the economy.

If in 2003, instead of starting a war, we had put our money into transforming our energy infrastructure to make it cleaner and more efficient, we could have spent less and been well on our way to ensuring a livable planet for our children. This complete transformation can be bought for less than 200 billion dollars. That's less than the cost of fighting in Iraq for two years.

Edwin Land, who invented the Polaroid camera, said that people who seem to have a new idea often have just stopped having an old idea. Amory Lovins, who worked for him, quoted Land as saying, "Invention is the sudden cessation of stupidity." Lovins is one of the most influential environmentalists around, because his aim is to make the improvements in the way we use energy more attractive by making them more profitable. Lovins, another MacArthur Foundation fellow, is attempting now to move American auto manufacturers in the direction of being smart people doing smart things, rather than smart people doing stupid things.

One reason why Lovins is keen to do this is that he has taken seriously the question of what the quality of life will be twenty or fifty or a hundred years from now for our children and their children if we don't change our ways. Lovins believes that our technologies are up to the task of keeping us as comfortable as we are now, even as we change our ways. As he wrote, "It will cost less to displace all of the oil that the United States now uses than it will cost to buy that oil." A conventional car uses most of its fuel moving steel, not people. That's because people don't weigh as much as steel. If we made our cars out of carbon fiber instead of steel, we'd save at least half the weight, use half the fuel, and the cars would cost the same to manufacture. (They would also be safer to drive, because carbon fiber can absorb more energy than steel.) Why not do it? Because we've been richly rewarded for making steel, using oil, and selling big vehicles. The nonlinear part of our brain is satisfied that our strategy is correct.

The rational part of the brain, however, can now understand that our strategy is actually one of Daniel Dennett's "good bargains" that has lapsed. And in any event, as Schrag pointed out in his lectures, within thirty to fifty years, how we use oil will have become a moot point, because we will have used up all the oil that the earth can provide. The same will most likely be true of natural gas as well. When my son Jonas is my age, his car, if he has one, will not run on oil. And frankly, no one has a very good idea of what it will run on, for all the alternatives, from hydrogen to ethanol, come with costs as intolerably high as those for oil and coal have been.

The oil companies themselves are beginning to recognize this. The chairman of the American Petroleum Institute, Mike Bowlin, said, "We've embarked on the beginning of the Last Days of the Age of Oil." The CEO of Texaco said, "The days of the traditional oil company are numbered." Most of our pollution, however, does not come from oil. It comes from generating electrical power by burning coal. Forty percent of that power is lost moving it across the country through wires. And a third of the rest of that electrical power is used for appliances that were invented for our convenience, not for our survival.

In order to change, we have to displace our old models of how the world works. While we blame corporations for environmental damage, they can do nothing without us. We consume what they make. Indeed, our culture and our economy are predicated on our vacation state of mind. But if our children are to survive, we must become smarter consumers. To go forward most cleverly, to get a better bargain, we have to know the rules that govern what can and cannot be done. So far I've been talking about how human systems work to shape our behavior and how understanding those rules can help us. But we are a product of nature, and unless we understand something about how nature works, we may do harm to ourselves in other ways as well.

The universe produced us. Its laws govern us. It is therefore a good idea to have some idea of the kind of universe it is and what those laws are. When we contemplate our own abilities to choose, it's important to remember that deep learning, the effort of thinking things through, changes the way we know everything else that we know. Every bit of new learning changes all of our knowledge. It alters the way we understand everything else in our world. It literally changes who we are.

Why does everything come with a cost? Why can't we have a free ride? Can we answer those questions?

Let's begin at the bottom, then, and look at some fundamental principles that make the universe do what it does. Since we are part of the universe, and since we have the capacity to understand these things for our own benefit, we're probably better off knowing than being controlled by forces that we don't understand.

I will start where I began, in my childhood, when I first wondered about the forces that shaped the world around me and asked what my place in it might be.

Nine

"The Earth Is Rotting"

When I was a boy of eight or nine, my father gave me my first job. I worked in his lab at the medical school where he taught. I swept the floor, washed the glassware, and took out the trash. It was there, in his lab, that I first had a glimpse inside the very workings of the universe, though I had no idea what I was seeing at the time. One day I was washing glassware, standing on a step stool to reach the sink, wearing an oversized lab coat and blue rubber gloves to protect me from the dichromate solution* we used to clean the glassware. I was rinsing a large bottle, holding it upside down, waiting for the water to slowly bubble out. My father came in, saw what I was doing, and took the bottle from me. He gave it a quick swirl, and a beautiful vortex formed inside the bottle. Within a second or two, it had organized itself so that there was a hole in its center, just like images I'd seen of hurricanes. The water drained out in seconds.

*This is potassium dichromate, one of the hexavalent chromium compounds made notorious in the movie *Erin Brockovich.* The solution also contains sulfuric acid.

"You see?" he said. "So simple. It's more organized. The universe likes a good trick."

And he went back to his Zeiss microscope and the small FM radio he'd built for himself (out of a new device called a transistor), which sat on the counter playing classical music.

From then on, I always drained bottles by giving them a swirl to create that vortex. I had no idea what caused it or why it worked the way it did to drain water so fast. As I would later come to understand, it was an example of how a natural system will sidestep the most direct route to solving a problem and choose a better path. It was an instance of a natural system having memory and forming what almost seems like a mental model. Of course, without mind, there can be nothing mental about it. But it was a template for behavior in which what had happened in the past influenced what happened in the future. At the time, I didn't understand any of it. Part of my education, as I came to see, required that I be baffled at first. Curiosity, confusion, learning, and at last understanding—that was what hope held out for me.

One of my favorite parts of our workday was lunch. At lunch, my father was all mine. There was a greasy spoon across from the Houston Medical Center, and we'd cross the asphalt parking lot, hot as a furnace at midday, and sit in the window of that diner eating cheeseburgers slathered in mustard and French fries drenched in ketchup, while the pile-driver sun streamed in. (This was before we came to count the cost of our eating habits.) It was at times like these that I could ask him all the questions I had stored up while he was working.

Shortly after he had shown me the proper lab technique for emptying a bottle, I asked him what made the vortex. I was just a kid, and I can't remember his exact words, but the conversation went something like this.

"It moves energy faster that way."

"How does it know?" I asked.

"It doesn't. It just happens through natural forces. Gravity is trying to pull the water down. As the water moves down, a partial vacuum forms in the bottle. Air tries to move up through the water to even out the air pressure."

"Yeah, so it just bubbles."

"Yes. Until you swirl it." He explained that once the molecules of water get going in a circle, a spiral forms, because gravity is still pulling them down. Air is still pushing up into the bottle, so it forms a sort of elongated bubble, and as the air bubble reaches the upper surface of the water in a bottle, a channel opens up with the water flowing around it. Once the channel is open, it won't close as long as water and air keep flowing. The vortex is essentially solving the problem of two kinds of molecules—gas and liquid—trying to get out of each other's way. They can't pass through each other, because of the Pauli exclusion principle (a term I didn't understand at the time). So they have to find a way around that problem. Once the channel is open, the water drains and the air fills the bottle very fast, because the molecules aren't getting in each other's way so much.

"Why?" I asked. "Why do they have to go anywhere at all?"

"Well, that's deep. There are a number of ways to answer that. One reason is gravity, obviously. And quantum mechanics, which involves the Pauli exclusion principle."

"What's that?"

"It's complicated. But one thing the Pauli exclusion principle means is that two things can't be in the same place at the same time. Another thing you have to consider is thermodynamics."

I'd heard of it but didn't know what it was.

"The second law of thermodynamics says that things spread out," he said. "That's a crude way of putting it." He pointed at the hot cup of coffee he was drinking. "If I leave that here, it'll be cold in a while. The temperature in the coffee and in the room will both converge on the same value."

"What's that got to do with the vortex?"

"The air pressure in the room and the air pressure in the bottle are trying to converge on the same value," he said. "And the vortex is a good trick for doing that."

"I still don't get it," I said.

"You like to play pool over at the Jenkinses' house, don't you?" he asked.

"Sure."

"How are the balls arranged when you start?"

"In a triangle."

"And you break the balls by shooting the cue ball into the triangle arrangement. Then what happens?"

"They go all over the table."

"Right. And if you knock the balls around, would you ever expect them to wind up in that triangle arrangement just by chance?"

"No."

"Why not?"

"I don't know. It just won't happen. You have to rack 'em."

"Exactly. And it takes work to rack them. Well, think of all the molecules in this room and in the coffee cup as pool balls—which of course, they're not. There's this invisible pool shark who's constantly knocking the balls around. He doesn't like neat arrangements like the triangle. So he'll do anything he can to mess them up."

"There's no pool shark," I said.

"Okay, there's not really a pool shark. But when one pool ball hits another, they both fly off in different directions. But they're both moving at different speeds now—different from the speeds they started at. The one that was moving is going more slowly, and the one that was sitting still is now moving. The moving one gave up some of its speed to the stationary one."

"Okay . . . so molecules are doing the same?"

"Right."

He explained that all molecules are moving around all the time, all at different speeds. Temperature is just a measure of the percentage of speedy versus slower molecules. The hot coffee cup has a greater percentage of speedy molecules than the air has; in everyday language, we call that hot. When one of those speedy molecules hits a slower molecule, the speedy one slows down a bit and the slow one speeds up a bit. The speedy molecules of the coffee give up some of their speed to slower air molecules. That means the cup is gradually giving up its extra energy to the air through heat. After a few trillion collisions, the average speed of all the molecules throughout the room and the cup will reach about the same value, and no significant change in temperature will happen after that.

"It won't go the other way," he added. "The cup will never get hot on its own, just like the pool table won't rack itself."

"I don't know what you're talking about, Dad."

"You will."

Some time in 1958, my father's department ordered the newest research tool of the biological sciences from Siemens in Germany: an electron microscope. We anticipated its arrival with the excitement you might reserve for a new baby. After some months, it arrived and was installed in a special room of its own. But by the end of 1958, my father was in a state of high agitation. The scope just sat there. No one showed the least interest in it.

My father scoured the burgeoning Houston Medical Center campus for someone who might help him figure out how to use it, because, as he explained to me, it promised to reveal the very workings of a living cell. For the first time, you could see right down to the finest structures. And it was driving him crazy that he could not look for himself. He had to know. Normally an easygoing, fun-loving fellow, he grew sullen and short-tempered. He couldn't sit still until he had his hands on that thing, and he was astonished and I think a bit embarrassed that none of his colleagues understood the importance of this opportunity.

By the spring of 1959, he had arranged for a grant so that he could go to Harvard University and learn the techniques of electron microscopy. That summer, my mother and brothers and I joined him there. I was eleven years old. While my father worked in darkness, I spent hours in the Boston science museum, watching an exhibit in which water would repeatedly freeze and thaw in a great glass aquarium. I could see the stunning phase transition water undergoes as it reaches 32 degrees (0 C). Liquid water looks the same from every angle. But as it begins to freeze, suddenly that symmetry is broken by crystals that look different from all angles. Decades later I'd learn that that same sort of phase transition played a role in how our universe formed.

When we returned to Houston, my father was eager to try out his newfound technical skills, and he couldn't wait to show me. (It seemed that I was the only other person there who even cared to look, though that would change soon enough.) I remember the first time I saw the gleaming Siemens scope in its clean room. It was tall, reaching to the ceiling (or so it seemed to me). It looked like a great steel espresso machine. A bundle of cables issued from its domed top, and dozens of knobs, lights, and switches adorned it. At the base was a thick glass window a few inches square, and above that, a binocular microscope for viewing the images on the phosphorescent screen inside. It was there, in the darkness, with the screen glowing green, that my father first showed me the inner workings of a cell. I was astonished to learn that the inside of a cell, which I'd thought of as an irreducible blob of jelly, was filled with molecular machinery, pumps and shafts, bearings and conveyor belts, assembly lines, all working at something. Of course, the tissue we viewed was dead and had been preserved so that it could withstand the powerful electron beam, but the structures we saw there were up to an astounding array of activities in real life: there were creatures walking to and fro, bearing great zeppelins of materials, and assembly lines putting together exotic creations like fantasy towers in a cityscape. The concept of protoplasm would not survive that revelation.

One Saturday at lunch, I was asking my father about the vortex in the bottle again, and he said that it produces entropy faster than the chaotic bubbling does. I had heard the term "entropy," but I didn't understand it. He said that in a system (such as the one that includes the bottle of water and the earth) where there is a reservoir of energy that is high relative to its surroundings (such as the water in the bottle held up above the sink), the second law of thermodynamics says that the energy won't stay concentrated there but will spread out as quickly and completely as possible. Once it spreads out, it can't do work anymore. And entropy is a measure of how much of the energy that was available to do work has been rendered unable to do work.

"That's a dumb thing to measure," I said.

"Well, temperature is a pretty dumb thing to measure, too, as far as that goes," he said.

"It tells me if I have to wear a coat or not."

"Yes, but that has to do with the ordinary world of everyday things. Entropy tells us more about how the universe works."

Any change in the world, he said, is attended by an increase in entropy. "Think back to the pool table," he said. "If the balls are racked, everything is neat and tidy. But you can't make them more racked. Any change you make is going to make them more messed up, and the entropy will increase. If you like, you can think of entropy as a measure of the change from orderly to disorderly. But that's not exactly it, either."

He seemed a bit confused about it himself and fell into contemplation. We ate in silence for a while, and then he asked me what the most likely arrangement of the pool balls on the table was. I said I didn't know what he meant by likely. He said, "If you just threw all the balls on the table, how would you expect it to look when they stopped moving? All in a neat triangle like when you rack them?"

"No."

"They'd never wind up that way?"

"No, I don't think so."

"That's another way of looking at this." He explained that things in nature tend to go to their most likely state. "If you rack for nine-ball, for example, you can tell me exactly where each ball is. They're racked in numerical order. So we'll call that a very orderly state. It's also a low entropy state. High order equals low entropy." Once you break the rack, he said, we'd have a very hard time predicting where any given ball is going to wind up. "That's a disorderly state, and it represents higher entropy." There are very few ways to rack the balls but many ways to spread them chaotically across the table. Nature chooses from the most likely states, which also happens to take things in the direction of greater entropy.

He hastened to add that the pool table analogy was limited. Then he started to write an equation on a napkin.

"Dad, I hate equations," I said.

"No, you don't."

"Yes, I do."

"Well, this one's easy."

He wrote that the change in entropy equals the amount of heat you put into the system divided by the temperature. He was right. I could see it. If you supply one unit of heat and the temperature is 100, that's $1/100$. If the temperature was lower—say it was 2—then you'd get $1/2$, which is a much larger number than $1/100$. The colder something is, the more entropy you can buy with a given amount of heat.

Entropy, he said, is all about systems, at least for our present understanding. (A substance can have what's called "residual entropy," but we don't need to get into that.) For our discussion here, there has to be a flow through a system, like a stream flowing into a millpond. As the water turns the wheel, it does work, produces heat, and increases entropy. When the water reaches the pond, it can no longer turn the wheel. So energy moves, work is done, and entropy is produced in that process. The mill, he said, uses mechanical energy. As the water comes down the stream, all the molecules are moving in the same direction. That's very orderly. Once they're in the millpond, they're all moving chaotically in every direction. That's disorderly. Entropy has increased.

He had gathered the check and left a tip on the table. He explained that you could do the same thing with chemical energy instead of mechanical energy. "For example, once your burn a log, you can't get the energy back into the log. It's a one-way process." The log represents an orderly state of high-energy molecules. The gas and heat that come out of the fire represent a disordered random state.

We were getting up to go back to work, and I felt that I was on the brink of understanding something important, something elemental about my world.

As we were leaving, almost as an aside, he commented, as if trying to puzzle it out for himself, that somehow or other, under the right circumstances, that process of moving energy around can create organized systems like the vortex that generates more entropy faster.

"But *why* does it happen?" I asked.

"The scientist doesn't ask why," he said. "He asks how."

I used to like to hang out in the gross anatomy lab and watch the medical students dissect cadavers. One day I was in there, watching an anatomist dissect the brachial plexus nerve out of the shoulder of a woman who would have been quite attractive if she had not been dead. I don't know where she came from; the school obtained bodies from all over the place. The anatomist, a fortyish man with gray hair at his temples, wearing a stained white lab coat, teased the nerve out from the surrounding flesh with a scalpel. It was the weekend, and he was just practicing his technique, he told me. No one else was in the lab, and the fluorescent lights were off over most of the rows and rows of stainless steel coffins. The brachial plexus nerve looked like a white piece of string. He remarked how beautiful it was and explained to me that if it were cut in a living person, it would render the arm useless. (Herophilus dissected live people in Alexandria in the third century BC to demonstrate such things.)

As I was taking in this beautiful complexity, I was suddenly struck by a thought. If the second law of thermodynamics says that everything has to go to equilibrium, to its most likely state, to spread and dissipate, then what power brought this creature together and made it dance and kiss and laugh and toast the new year? For that matter, if ever-increasing entropy worked to scatter everything, what gathered together the materials and organized them to make this lady grow up in such a complex and orderly fashion? Surely life was breaking this famous second law.

The next time I was at lunch with my father, I challenged him with that thought. "There must be some big piece of the puzzle missing," I said, "some other law that said, let flowers grow, let ladies sing—or something. And anyway"—said with what must have seemed a perfectly adolescent fervor—"if this law's so special, why isn't it called the first law of thermodynamics? Why is it second? Maybe there's a better law."

I remember him laughing at that. "That's the ticket," he said. "You've got to get mad when you don't know things, and then go storming out into the world demanding answers." But when I pressed him on the question of life, he said, "Well, it's a mystery." Then he added with a smile, "But we're working on it."

I felt somehow disappointed. I had always relied on him to have the answer to every question I could think of. But that day, faced with the mystery of life, he at last admitted that there was something we couldn't simply look up. The question was so big that we didn't even know how to approach it. When, in my childish impatience, I demanded answers from my father, he told me what he'd told his students on many occasions: "We teach MDs answers. We teach PhDs questions."

So I began the long process of attempting to learn. What I began to learn was that finding good questions to ask is sometimes more important than trying to get a solid grip on definite answers. And at my father's side I came to understand that learning is a bit like hunting must have been long ago. Tasty little creatures will not just leap out of the forest and into my arms for the pleasure of feeding me. I have to work them out of their burrows by my own strength and cleverness or otherwise go hungry. As with the necessity of eating, I have to do it again and again. And with each meal, I can learn to love new and more complex flavors.

Eventually I'd learn that I was not alone in wondering about this question of how life could be reconciled with the second law of thermodynamics and the tendency everywhere to produce more entropy. I'd learn, too, that both my father and I were harboring some misconceptions about the whole business of thermodynamics and entropy and energy. It's no wonder. They're pretty slippery concepts to discuss without mathematics. One reason we have trouble understanding entropy (or quantum mechanics, for that matter) is that it is difficult to create a mental model for it. Where there's no mental model, there is nothing that the human mind can hold. For those who can master it, mathematics provides a new kind of mental model to hold such abstractions—or failing that, at least a stepwise procedure. On the question of thermodynamics and life, however, I'd be lucky enough to live through a time in history when some pretty good answers to my questions were being proposed. It all had to do with finding out the cost of the things that we—and nature—do.

Since researching this book, my own life has become a bit surreal in ways that I could not have imagined. Once you start thinking about science and questioning the cost of things, the world begins to look very different. For example, take my own little neighborhood. I live in a quiet suburb. It is American and ordinary in every way. I'd never really given it that much thought. The houses are neatly lined up, all designed by the same man just after World War Two. My neighbors put out holiday decorations and political campaign signs in the appropriate seasons. They drive the large domestic trucks that people refer to by the somewhat Orwellian term "sport utility vehicles." (While I once may have been able to imagine driving as a sport, it's become increasingly difficult to do so, and the utility of those vehicles has become questionable at best.) During the summer, the neighborhood echoes all day long with the roar of lawn services as they tend the grass and flowerbeds, filling the air with smoke. I reflect that nice lawns and gardens for cheap is the gift. Noise and smoke are the cost.

In this apparently normal scene, I began experiencing shocks, visions, epiphanies. One day, I was walking up my driveway toward the alley carrying two plastic garbage bags full of trash, and I stopped in mid-stride with this thought: What on earth am I doing? I'm carrying these two heavy bags full of manufactured goods, and I'm going to place them in this container in back of my house. Tomorrow a big truck will come roaring up the alley, belching smoke, and will empty the contents of that container into its belly and go away.

All up and down the block now, I notice the garbage cans in ranks like open mouths, waiting to be fed. What is the nature of this beast that I'm feeding? Is it a god or a demon? Why must we all feed it so much perfectly good stuff? For most of what I throw out is not inherently bad or useless. Some of it, to be sure, is eggshells and onion skins—and even those could go into a compost heap, if I had one. But most of the space in these bags is taken up by beautiful creations, glassine boxes that once contained olives which came to me all the way from Spain, or clusters of tiny cherry tomatoes grown in New Zealand. There's a waxed cardboard vessel, brightly decorated, that once contained the juice of a tropical fruit that doesn't grow anywhere within hundreds of miles of my home. There are gleaming coppery tins that contained small delightfully seasoned fish, which once swam in a sea halfway across the globe off the coast of Norway. Inside one of the bags is an oblong box decorated with artfully printed, four-color photographs of food. It had contained a glass vial of an exotic pepper sauce from Avery Island, Louisiana. I'd scarcely glanced at it a couple of days before when I took out the bottle of Tabasco sauce and threw away the box.

The bags I carried that day contained what might have been a virtual treasure trove had it been given to a person, say, 50,000 years ago. It would have been a miracle of astonishing proportions just to glimpse these unimaginable things.

Now, all up and down my street, people like me were taking these gifts and feeding them to the creature with a million mouths, so that the trucks could come and take them to the outskirts of town and dump them in a huge heap, where all that beauty, all that craft, would lie, inert, useless, the collective effort of millions of people, their lives expended in making those gifts, now thrown out so that they could make that much more for next week's ritual sacrifice when the trucks would come again. As I stood there, stunned by this concept, I realized that I was vouchsafed my comfortable way of life only by dint of how well I played the role of the garbage producer. If I stopped—if we all stopped—the system would collapse. The great wheel that turns out all this artistry would grind to a halt. The goods would grow dusty on the shelves. The highways would grow silent. The skies would clear of contrails, and the great screech of industry would fade away.

These intrusive thoughts came to me unbidden as I learned more and more about the world and how it works—how it has rules that can't be broken, costs that must be paid. I remember standing in the supermarket one day and turning around and around at the wonder of what I was seeing, this astonishing feat. It was a dark and wintry day outside, and the store was heated to a comfortable temperature and blasted with fluorescent light—so bright, in fact, that we could have done brain surgery there if the need suddenly arose. I stood in the dairy aisle, and on one side was an open refrigerated chest longer than half a city block filled with cheese and milk and eggs and juice. On the other side of the aisle were ranks of freezers filled with ice cream and ready-made breakfast foods. This was the arrangement: the entire store was being heated by one set of machines against the wintry weather outside, while another part of the store was being cooled by another set of machines against the warm temperature created by the first set of machines. (Most of us have this same arrangement in our homes on a smaller scale.) This seemed to me a puzzle within a puzzle. It cannot be an accident that this process is so wasteful. It seemed such a mad enterprise, but I knew that there must be a deep and secret law hidden here.

And that's to say nothing about the products that this clever arrangement was designed to offer. They are a wonder of human imagination and ingenuity. Velveeta, ready-to-eat Philadelphia brand chocolate cheesecake filling, Cheez Whiz, Jell-O and custards and puddings of all sorts, already cooked and cooled. Butter flown in from Ireland, and products that look like butter but that cannot be explained without a PhD in organic chemistry and as a consequence have names like I Can't Believe It's Not Butter! and Egg Beaters, and Bob Evans Snackwiches, and Dannon Frusion Smoothies and Fizzix Fizzy Yogurt Snack, drinkable yogurt and portable yogurt. Whipping cream and heavy cream and sour cream and Reddi Whip, and individual slices of American pasteurized "process cheese food," each wrapped in its own glassine envelope. Juices—orange, grape, passion fruit, mango, grapefruit, pineapple, strawberry, banana . . . Eggo pancakes and French toast and waffles and toaster strudel and Edwards Oreo Cream Pie to top it off. That's just for breakfast. It doesn't even take into account the vast selection of paper products to be used and quickly thrown away, a universe of cleaning products, a citadel of breakfast cereal, a wall of soft drinks, a sea of bottled water, wine, beer, a bakery, a butcher, a florist, a bank, and a greengrocer selling fruits and vegetables that grew as far away as Australia, Japan, and China. The average bite of food I eat travels 1,500 miles to reach my mouth, and each calorie I consume from this supermarket costs fifteen calories to make. What I eat could feed fifteen people.

All of this plenty, this profusion of good things, seemed arrayed there for no purpose other than my enjoyment. What an astounding amount of effort had gone into pleasing me, I thought. I realized that this colorful profusion must be an echo of the primeval rain forest, the dazzling variety and sweetness of the fruits our ancient ancestors found in the beautiful (if dangerous) world where they lived. But here it all was, laid out for me with the danger apparently removed. I must be someone really special to deserve this treatment. I must be . . . God.

But of course, I'm not. I'm flesh. And the danger has not really been removed. It has simply been hidden. These products are here because I have to eat, and to the extent that clever people can make it easier for me to eat, they can separate me more readily from my money. No animal refuses a more convenient supply of food. All the while, disembodied voices in this illuminated grotto wheedle on with sham goodwill, cajoling me to buy more and more and more, "While supplies last." The supplies could run out sooner than they think.

As I left the store, I noticed a shelf on which Slim-Fast was being sold right next to Ensure—the first product to keep us from getting too fat ("controls hunger up to 4 hours"), the second to prevent us from wasting away. I left wondering: Can we get away with this? Is it allowed? What are the rules by which we operate? Is there some deep message in the remarkable profusion I see here? If this supermarket is part of the universe—which it clearly is—what set of rules gave rise to it? Because, as with the arrow I had seen in Copper Canyon, it might not seem to be such a great mystery to say that a man makes a supermarket. But it is a great mystery to say that nature makes a man who makes a supermarket. Exactly how does the big bang, some 10 or 20 million years ago, go about producing Fruity Cheerios? Why would nature do such a thing, and by what means? Who organized this elaborate system? Is it possible that such a marvel could organize itself?

If you take apart a traditional wind-up clock, you may have a moment of revelation in which you grasp, completely and precisely, how it does what it does. There has been no such moment when it comes to trying to take apart what scientists refer to as self-organizing systems.

A good example of a self-organizing system is the sand pile. If you sift sand from above and let it pile up for a while, it will eventually make a little mountain and grow no taller. Add more sand, and the pile will start to collapse. This is where it gets interesting. Sometimes a few grains tumble. At other times the whole pile seems to undergo a tremendous avalanche.* Avalanches of all sizes occur. It's not random, either. If you measure them, they make an orderly pattern.

A key to understanding this sort of system lies in realizing that there is a flow. The system is processing something. Sand is added at one end and it leaves the pile at the other end by means of the avalanches. Falling sand has energy (the energy of falling) and that energy is transformed by the system through heat from friction as the sand grains grind against one another. Two elementary features of the universe are opposing each other to create this system: the force of gravity, which pulls things together, and the Pauli exclusion principle, which keeps them apart. (As mentioned, for our purposes here, the Pauli exclusion principle can be thought of as just a fancy way of saying that two things can't be in the same place at the same time.) In between the two ends of the system—input and output—is order, information, and complex behavior. It is a pattern, some sort of information, a message from the universe. We don't know what the information coming out of these spontaneous structures means. In fact, we have only recently begun to notice them at all.

*A real sand pile does not work in precisely this way. The experiment was done with a computer simulation. A pile of rice grains works better, due to differences in size, shape, friction, and density. But "sand pile" sounds better than "rice grain pile."

The vortex in a bottle is a self-organizing system, too. Falling water, encouraged by a swirling motion, results in a complex, ordered structure that transports energy. When a few water molecules get going in a circular pattern, they bump into others and move them in a circle, too. The circle is being drawn downward by gravity, so it turns into a shape like a coil. The more molecules that get going that way, the more new ones are pushed, until a kind of stampede effect happens, and the whole system is wrapped up in the energy dance of the vortex. The tiny interactions of the little particles, by their very number, begin to produce complex outcomes. In a sudden, unexpected way, the whole system changes its character from chaos to order. A system like this has memory. Its present state depends on what happened in the past in an important way. When I swirl the bottle, it makes the water go in a circle. The water molecules have momentum, which just means that they keep on going when I push them. But momentum is a kind of memory. It projects into the future a motion that was started in the past. The motion I give to one molecule now predicts the motion of another molecule down the road. It's a very rudimentary kind of memory. But it is memory.

Scientists call these systems self-organizing, because no one plans or designs them, and because the effects we see can't be predicted by knowing about the individual parts of the system. Neither is there a clear definition of the system. Drop one grain of sand and you have nothing. Drop another, and it's the same. But at an ill-defined boundary, things start to change, and they change very fast. Suddenly you have the self-organizing system. We can watch the system develop, but there's little that we can say that will illuminate its interior workings. We can't take it apart and see it tick. No water molecule holds the key to the vortex. No sand grain can explain the avalanches of the pile.

In that sense, it's much like looking at elementary subatomic particles. You look and look, closer and closer, and the more you look, the less you understand. Life itself cannot be stripped down to its essential elements, either. No single molecule is alive, including DNA. The process itself is alive. The molecules that make up your body constantly depart, and new ones take their place. Cells die and new ones arise. Somehow through all this, you remain. In a few years, almost nothing is left of the body you had when you were born. And yet a coherent self holds its place like a stable eddy in a stream—a self defined by what's inside against what's outside. Life passes through us and then ultimately casts us off as it goes on. More and more scientists are coming to believe that life itself is—or is the result of—a self-organizing process. Or perhaps many self-organizing processes that interact.

Because these systems have memory, they also evolve. No two vortices will be the same. No two sand piles will have the same collapses. No two thunderstorms, no two lightning bolts, no two stars or chrysanthemums will ever turn out exactly alike, though the same process of creating them takes place each time.

The stock market is another example of a self-organizing system. It's made up of a lot of individual agents—people—each of whom is bound by simple rules: you can buy or you can sell. That's really all you can do. Yet the results are remarkable. Like the sand pile, the stock market is beset by seismic events, booms and busts, that no one can predict. Even when there is no large event, no one can predict the fluctuations in price of either a given stock or the system as a whole. Nevertheless, there are fluctuations of every size, and they follow a pattern much like that of avalanches. You can predict the frequency with which each size of avalanche will occur, even if you can't pinpoint when any given one will happen. The events and structure that arise in such a system are sometimes referred to as "emergent," because they simply emerge without any known cause. Anytime the stock market crashes and you read an explanation of why it crashed, you can be sure of only one thing: the explanation is incomplete at best and probably dead wrong. Events like stock market crashes happen at all scales—big and small—and, frankly, no one knows why.

Does everyone at the New York Stock Exchange, on a prearranged signal, send a text message to everyone else and say, "Sell your stocks now!" No, I don't think so. The crash isn't caused by an event. It's a characteristic of the system.

Don't feel bad if you find self-organizing systems a bit confusing. Everyone does. The Santa Fe Institute was founded in part to study these kinds of systems. I asked Doyne Farmer, a professor at SFI and one of the pioneers in the field of self-organizing systems, what progress had been made in the last twenty years in understanding why these effects arise, and he told me, "Not much."

You may be wondering at this point what self-organizing systems have to do with the variety of breakfast cereals in the supermarket. In one sense, they are directly relevant, because economies and markets are self-organizing systems themselves and give rise to behavior every bit as weird as the patterns of earthquakes. The supermarket is a result of a lot of different agents operating according to some fairly simple rules. But the point I wish to chip away at is a deeper one. I'm trying to dig down to some fundamental principles of the universe itself—from which we arose somehow—in order to see if our origins can help explain our odd behavior. And that means first understanding a few things about how the universe works. Along the way, I'm trying to understand, too, the costs of our behavior. This business of self-organization appears to be terribly important, so let's explore it a bit further.

As I look out my window at the trees, I see that they have a branching structure that makes it difficult for me to tell how big any part of the tree really is without something to compare it to. Little clusters of branches look pretty much like the bigger ones. If I look up at the sky and examine the puffy clouds floating there, I realize that I can't tell how far away they are. Am I seeing a small cloud close up or a big cloud far away? There's no way to tell without putting an object up there to show the scale. These types of structures, which look the same at all scales, are found abundantly in nature. They're called fractals.

The mathematician Benoît B. Mandelbrot, widely regarded as the father of fractal geometry, defined a fractal as "a rough or fragmented geometric shape that can be subdivided in parts, each of which is (at least approximately) a reduced-size copy of the whole." If you use mathematics to interpret a self-organizing system in a geometrical way, you can produce all sorts of artificial fractals. But nature loves fractals. Your own circulatory system looks like a fractal (though, if you want to split hairs, it's not exactly). I'll call it an "almost fractal," a near approximation.* This quasi-fractal structure, with identical features at many scales, breaks down below the level of the capillaries.

The circulatory system does, however, appear to be an expression of a self-organizing system, a network of remarkable complexity that arose on its own through a number of agents (chemicals) operating according to a set of not-so-very-complex rules (the making of chemical bonds, which means the exchange of electrons, which means moving energy from one place to another). Through that process, the circulatory system evolved a very good shape for delivering blood and nutrients to the cells of the body and for removing waste. In order to stay alive, animals need to get various materials to and from cells throughout the organism. The circulatory and respiratory structures, which look like fractals, are a reflection of what appears to be the best solution to this problem: a branching, hierarchical network that efficiently fills space while leaving room for the cells that it's supplying. Fractals are very economical shapes for filling space with a lot of structure. The fact that plant and animal vascular systems look a lot alike, even though they don't work alike, comes from the process of finding the best way to fill space. It's difficult to imagine that logical, stepwise engineering could improve upon it.

Those networks should require the least energy to do their job,

*The stock market has what's known as a "multifractal" structure.

while at the same time processing as much energy as fast as possible, given the rate at which cells burn fuel. That makes sense for living networks, which move energy around for cells to use, but it also illustrates a key feature of many natural systems, living or not. Fractal or near-fractal shapes are common in nature because they work well. The shape of a tree is good at transporting fluids and gathering sunlight. The brain's structure is a good way to create rich networks of neurons. Cut a cauliflower in half, and you will see that its interior looks like a beautiful fractal. Its exterior looks like a brain, which resembles billowing cloud formations. This isn't a curious coincidence. It's a result of the laws of physics that control how energy can be moved around, and the fact that it has to move around in the three-dimensional space where we live. Because the second law of thermodynamics says that energy must move, the shapes we see tend to represent good ideas for moving energy across a particular space. If we explored the coast of England, for example, and found that its shape turned out to be a good way to dissipate energy (say, between the land and sea), that would not surprise me at all. Such structures and systems may not achieve some sort of maximum ideal in processing energy, but they tend in that direction. They do what the vortex in a bottle was doing. Once the flow gets going, it tends to reinforce itself.

Mandelbrot used the coast of England to illustrate the idea of fractals. The coast has bays and inlets of all sizes, so that it always looks similar no matter what scale you choose. It looks similar whether you are viewing it from space or from a few feet above it or even at the microscopic level. None of the inlets will turn out to be larger than England or smaller than a molecule, but between the extremes there will be no typical size. Many things that we find pleasing to the senses also have a quasi-fractal character. Notre Dame cathedral, for example, looks complex both from a distance and up close, where you see all those odd little figures and designs. A researcher at IBM named Richard Voss showed that the tones in music vary in a fractal pattern.

One of the reasons I became so interested in self-organizing systems was that they seemed to fly in the face of what I thought I knew about entropy and the second law of thermodynamics. Things were supposed to get more messed up, weren't they? So how can they organize themselves? Why is this galaxy, known as M–74, so beautifully organized?*

If the universe is a stage set for the production of entropy, a game of probability run by an invisible pool shark, then doesn't it seem more likely that the big bang, as the beginning of the universe is called, would have resulted not in what we see around us but in a random gas spread out evenly everywhere, or perhaps in something with a simple, rigid structure, like a crystal?

I sometimes find it useful to think of the universe in terms of a casino. Casinos exist because the most likely events occur more often than the less likely ones. That's what "more likely" means. So if you play the game of craps, you quickly discover that the rules are arranged around the fact that seven is the most likely number to appear. There are six ways to make seven with two dice. Every other number is less likely to come up. One of the implications of the second law of thermodynamics is that where there are more pos-

*Credit: NASA, ESA, and the Hubble Heritage (STScI / AURA)- ESA / Hubble Collaboration. Acknowledgment: R. Chandar (University of Toledo) and J. Miller (University of Michigan).

sible paths to a given state, that is the state you are most likely to see. So I would expect to see a universe that looks like a random gas. It is possible that all the air in my study will suddenly rush to one corner, suffocating me (and heating up enough to set the house on fire). But I don't lie awake at night worrying about it, and I'm not in the market for a low entropy detector, either.

The games in a casino are designed to give the house a statistical edge, so that over a vast number of plays, the house will inevitably win. Even when you win, someone else in the casino is losing. They're losing more than you're winning. If you play long enough, you'll always lose at least as much as the percentage known as the house edge. So we can view our world—the universe—as a casino, in which we pay a price to play the game. We can build order here and appear to win. But the price for reducing entropy in our little corner is that we create even more entropy everywhere else.

This concept applies directly to our everyday lives. We tend to create order of some sort in our little corner of the world and then declare that we have won the game without taking into account the costs. We delude ourselves. For every cell phone we make, every iPod or computer or television set, for every beautiful and useful invention we create, we also create toxic waste dumps in Asia and Africa, where little children are playing out their lives right now. For example, children in Accra, the capital of Ghana, spend all day stirring fires to burn the insulation off of wiring, because the copper will bring a few dollars. Their lungs are filled with smoke containing potentially lethal doses of lead, mercury, arsenic, cadmium, barium, hexavalent chromium, dioxins, and beryllium. We throw away about 40 million computers, 25 million television sets, and 98 million cell phones each year. Most of that 50 million tons of electronic waste is sold to brokers who send it to those children for processing. When the children are done scavenging the saleable components, they throw the rest into the sea.

So yes, we have created a remarkable and beautiful enclave for ourselves here. But the cost is very high. Toxic waste dumps are not entropy, but the concept of entropy can serve as a useful analogy for these costs. The vast profusion of wonderful things to eat in my local supermarket represents a huge win for me in the casino of life. But it also represents a huge loss for someone else, someone who remains, for the most part, conveniently out of my sight. The peril of Lucy's life has not been vanquished by our achievements. It has been shipped overseas.

We can think of entropy, then, as the house edge. The universe extracts its toll through the production of entropy. Entropy is no more difficult to understand than money. Both are somewhat abstract concepts, yet we deal with money every day. We deal with entropy, too, as the cost of being here. We're just not used to thinking about it. But entropy is a price we pay for everything we do. And it is, in fact, the reason that we can be here at all. Entropy is the exhaust pipe of work.

Think of it this way: money represents time spent living. If I work eight hours and collect my pay, each time I spend a dollar, that represents a part of that eight hours. I'm paying, literally, with my life. I can never go back and retrieve the time I spent earning that dollar. My time, my life, is a useful quantity to me, like energy. I have the dollar, which is a measure of my life. Once I spend it, I can't unspend it. That bit of my time on Earth is gone.

Entropy production is a similar process. Each time you do anything, you do something to a bit of energy and produce a bit more entropy. The reason that more people play the slot machines than, say, baccarat, is that a given amount of money will buy you more time playing the slots. Baccarat is a high-stakes game. Think of it as a hot game. High-rollers from Japan and the Middle East play it. They bet hundreds of thousands, sometimes millions, to play. A hot game takes more money to play than a cool and casual game like a slot machine. The same is true in the universe. The reason that heat flows from hot to cold is that the same amount of heat energy will buy more entropy in a cold area than in a hot area. If you enter a hot baccarat game with $100,000, you won't last long. If you go to the 25-cent slots with that same amount, you could play for a very long time indeed. Entropy equals the amount of heat (money) you put into a system divided by its temperature (how hot the game is).

Another way to think about this is to recognize that this book, which contains about half a million characters, is in an extremely unlikely state. If you took the twenty-six letters of the alphabet and the ten digits of our counting system and spewed them out onto these pages at random, there is a chance that you'd get this book. But the chance is so small that you might have to wait longer than the age of the universe—and then some—before it happened. What you'd see the rest of the time is what's most likely: random arrangements.

So how did this book come to exist in the face of such universal laws? I'm a bit embarrassed to say: I'm afraid I had to produce a tremendous amount of entropy to bring it into existence. I burned through a huge amount of energy, ranging from the food I ate to the electricity that runs my computer to the reams and reams of paper on which I printed it, revised it, and printed it again and again. Before it reached your hands, it involved the squandering of far more energy as it was printed, manufactured, bound, and shipped around the world. Coal, oil, and forests were consumed in order to make this book. Whatever its value as reading material, whatever its internal order, it represents a great disordering influence in the universe at large.

Because life is in such an ordered state, it superficially seems to represent one of those impossible coincidences, the equivalent of spewing out random letters and winding up with this book. Or the equivalent of breaking the balls on a pool table and having them wind up back in a neat triangle. Or, perhaps as unlikely, having the big bang create the perfect spiral galaxy, M–74, or my local supermarket.

Erwin Schrödinger, one of the fathers of quantum theory, was one of the first scientists to pit the second law of thermodynamics against life and to ask why nature is complex instead of simple if everything tends to move toward the statistically most likely state. (Seventy years earlier, Ludwig Boltzmann had pointed out that photosynthesis seems to be a process that is contrary to the second law of thermodynamics, but his idea went nowhere at the time.) He must have realized that the complex structures we see at every scale, like M–74 and the vortex in a bottle, are good at allowing energy to move and change, and so they are also good for making more entropy.

When we say that something makes sense, we mean that we can make it accord with the physical world, which we perceive through our senses. But that is a narrow view. Life, at first glance, appears nonsensical according to the rules of physics and chemistry. The materials in a living cell move about randomly, but the genetic code does not allow normal chemical reactions to take place at random as they would in nonliving matter. DNA actually chooses very specific chemical reactions in defiance, it would seem, of the normal rules of chemistry. DNA selects chemical reactions that, taken in isolation, go against the normal thermodynamic direction of dispersing and producing more entropy. Life chooses high-energy molecules to accomplish that. Living things decrease entropy within themselves. But they produce more entropy in their surroundings. Their orderly activities are just as costly as our winning at the casino is. Just as costly as this book.

Systems that spontaneously organize themselves were first described in detail by the Nobel laureate Lars Onsager at Yale in 1931. Alfred Lotka, a Ukrainian statistician and chemist, called them dissipative systems, because they dissipate energy, which makes that energy useless for doing work. You can think of the order that dissipative systems produce as lying somewhere between the giant crystal universe and the random gas universe. On the one hand is perfect order, on the other complete chaos. Why do such systems organize themselves? We don't really know. But they seem to be awfully good at doing the work of the second law of thermodynamics.

It is through the contemplation of such mysteries that I have come to my strange view of the world where we live and the universe at large. Self-organizing systems are connected to the supermarket where I shop and the garbage that I throw out for many reasons, but the one that seems most urgent at the moment is that self-organizing systems can be very delicately balanced, and therefore can be very sensitive in the way they behave. For example, we're all familiar with the annual drama involved in watching a hurricane move across the ocean. The hurricane itself is very stable. It would be hard to make it go away. But the exact way it's going to behave is completely unpredictable.

The weather of the earth is itself a big self-organizing system. Through the use of computer models, scientists make predictions about what effects warming up the earth will have. But, as Daniel Schrag has pointed out, those models are crude, and the behavior of the earth itself is far more complex. We are, he said, in for some big surprises. One of them seems to be that the ice is melting a lot faster than we thought it would. Various pollutants and greenhouse gases contribute to warming the earth, but carbon dioxide accounts for about 70 percent of that effect. And those pollutants are caused by my way of life. They are caused by the vast system of my civilization, which I experience only as a cornucopia of delightful things to do and eat and the need to throw out all sorts of manufactured goods every day. It is this system that's causing the problem. But I have a hard time seeing the problem, because this system ensures that I'm having too much fun to even think about it. How can I think about global warming when I have to rush home, order a pizza, and watch *Dancing with the Stars*? This is the grave danger that has been concealed in the supermarket and the garbage truck that is destined to receive all that bounty. We solved Lucy's problem, it appears. We vanquished our predators. But that's the illusion over here in this little corner, even as the costs of our achievements are mounting out of sight. The sleight of hand of marketing gives me the cell phone while cleverly concealing the poison at its core. (To name just one, computer chips contain arsenic.)

Our view of the world, the way we think, had its origins in the great rain forests that circled the earth tens of millions of years ago. The mental models and behavioral scripts that direct our behavior are given much of their shape by those origins. When we lived in small family groups of fifty or 150 or in tribal groups of a few thousand people, sharing a common language, our narrow view was sufficient to the task of survival. There simply weren't enough of us to disrupt the self-organizing system that is the earth. Later, as we grew more powerful and developed agriculture, we could disrupt a local ecology, create a local desert out of fertile land and not notice that we'd done it. We'd simply move on. There was still much room. But now there are too many of us who want to live the way I live. The room has run out. The self-organizing system of our weather is already wobbling erratically in its motion. The refusal of my country to accept limits on carbon emissions at the United Nations conference on climate change in Bali in 2007 is an ominous sign that nothing will change until it's too late.

Venus is a planet that shows us what a runaway greenhouse effect can look like. Its atmosphere is almost all carbon dioxide. The surface temperature is around 854 degrees (460 C). We have a small window through which to view what the earth might be like if we push carbon levels a bit higher. As mentioned, when the Industrial Revolution started, the atmosphere had about 280 parts per million of carbon in it, and in the preceding 650,000 years the level had never been above that. (The oceans and plant life on Earth are natural carbon sinks that siphon off the natural carbon produced by such things as volcanoes and rotting organic matter.) Today we have around 380 parts per million of carbon in the atmosphere. The last time it was that high was 35 million years ago, when palm trees grew in Wyoming and crocodiles lived in the Arctic. Antarctica was a pine forest, and there was no ice anywhere on the earth. Carbon had peaked at about 500 parts per million. If all the people on earth stop putting carbon into the atmosphere today, the level will still rise to 500 before it starts to come back down again. Then it will take another 2,000 years for the ocean to reabsorb all the carbon we've already emitted.

There is no longer any doubt about what is happening. Go to Peru, go to Kilimanjaro, go to Mendenhal glacier in Juneau, Alaska. The ice is going away. Snowmelt in the Sierra Nevada is the water supply for all of California. California is the source of two-thirds of all the fruit, vegetables, and nuts that I find in my supermarket. The snow has already started to melt early, and there have already been floods. Once the snow fails, there will be nothing to melt in the summers and hence no water. And that's just one state in one small place in the big picture.

What about the ice disappearing from the Arctic tundra? Do we really care that someone will no longer be able to hunt seals or polar bears? Maybe that has no direct effect on my way of life. But the tundra happens to contain an enormous reservoir of organic material, accumulated over hundreds of thousands of years and conveniently frozen so that it doesn't rot. Melting the tundra will open up that vast morgue of material, and bacteria will quickly consume it all and release all that carbon into the atmosphere. If that happens, it won't much matter what we do, because even if we emit zero carbon after that, we'll be unable to reverse the trend. So it seems that the rules of this casino universe led to life and the existence of life led, through evolution, to my particular way of life, which in turn led to this unpleasant predicament. I find this idea very puzzling. Are we just the victims of a trick of nature? Or do we have anything to say about it?

One afternoon while visiting the Santa Fe Institute, I heard the announcement for tea and went down the hall to the kitchen. Tea-time at SFI is a tradition going back to its beginnings, a way of getting everyone away from their work to talk. Geoffrey West, the president of SFI, was by the coffee percolator talking with the physicist Eric Smith when I walked in. Murray Gell-Mann was talking to Cormac McCarthy. Murray, as everyone at SFI calls him, is one of the most prominent scientists in the world today. He won the 1969 Nobel Prize in physics for his work on elementary particles. Born in 1929, he is a man of small stature and a busy mind that appears to take in and quickly analyze everything. When I met first him, he said, "Gonzales, eh? That has something to do with war." He was referring to the origins of my name. (It comes from a Visigoth word meaning "battle.")

When Cormac drifted away from his conversation, I approached Murray and asked if he would agree that life arises because its structure does a good job of moving energy under the circumstances and given the materials found on earth. He said, "The earth is rotting, and life is the waste." I puzzled over this strange statement for the longest time. To have a big unanswered question is a joy, as hunger is a joy before a meal. It would take me several years (and the rest of this book) to begin to comprehend what he meant. And when I did begin to understand his statement, I thought to myself: If that's true, then what is intelligence? What's consciousness? If that is our view, then how is it that we can have a view at all? And what is a view?

Living things are systems in which energy goes in one end (sunlight, food, chemicals). The energy lingers long enough to create order, structure, and information (such as feathers for birds or flowers for plants). That structure decreases entropy inside itself to allow a greater flow of energy and a greater increase in entropy outside itself. Then the transformed energy (some in the form of matter) flows out the other end. In the case of most animals, there is still energy left to transport at that point, so another species (an insect, a bacterium) takes up the flow. We are only just beginning to open our eyes to these types of systems, and much remains to be understood. But somehow, in the same way that photons and electrons seem to know where they're going, organic chemicals seem to know how to make life. They seem to know that life is a very good trick for making energy useless. Science has not yet figured out how that can be so.

There are only a few types of elementary particles in the universe, and yet they conspire to create the complex world we see around us, including us, even as we observe our world by using those particles. When we see, we are literally experiencing the universe looking at itself.

Ten

The Cosmic Cheat Sheet

I took Murray's cryptic statement—"The earth is rotting, and life is the waste"—as more or less an affirmative answer to my question about whether life exists because it is a good structure for moving energy. I wanted to pursue it with him, but he's a busy scientist with many questions already on his mind. Still, I felt that contained in his answer was a secret that might take me all the way back to the days in my father's lab and the vortex in a bottle. It had to do with the second law of thermodynamics, with the production of entropy, the movement of energy, and with the way that a natural system will sometimes sidestep the most direct path in order to achieve a better path. But before I could fully understand all this, I'd have to take a side path myself and learn more about energy and how it creates and uses structures, and most especially more about life itself.

Though we often live our lives as if we make the rules, we have been operating for millions of years by immutable natural laws and will either survive or perish by them. We now have a crucial advantage over all previous versions of ourselves, such as Lucy and *Homo erectus*. We have the ability to reflect on what we do and thereby earn the opportunity to change. By this light, then, we may actually see what we are doing instead of groping in the dark, as our ancestors were condemned to do. Understanding ourselves, truly believing that we are part of a natural system, requires forging a new mental model of our world and our connection to it, for we only see, we only believe, those things for which we have mental models. Our old mental models are rugged and hard to change. For me that meant that I had to go all the way down into the mechanics of life to see for myself. And so in my search for answers, I went down into the earth itself to glimpse the raw beginnings. From there I would gradually rise up and try to assemble a new understanding for myself.

I drove to South Dakota and stayed the night in Hill City. At first light, I made my way south along a winding road through Custer State Park. I could see Cathedral Spires in the distance, leaping into decks of pink clouds against a blue-gray sky. The moon set through the pines like a faceted ivory lens. It was fall, and the air was full of wild red and gold leaves, jerking in the wind like broken butterflies.

Rod Horrocks, an intense man of medium build with brown hair, blue eyes, and an easy laugh, met me at the visitor center. He was in charge of mapping Wind Cave, which is called the fifth largest cave in the world. But as Horrocks told me when I arrived, "Nobody will ever find the end of this cave."

Horrocks discovered caving while on a trip with his father at the age of seven and, by his own admission, he hasn't shut up about it since. ("A serious caver never says 'spelunking,' " he warned me.) There wasn't a hole he wouldn't scramble into or an underground feature that he wouldn't stop to explain. I didn't share Horrocks's passion. I don't much like enclosed spaces. But there were things in there that I needed to see in my search for understanding. This was a place where the great grinding machinery of the universe was at work. Something momentous had happened here. The earth was ripped up as the universe wrote its messages on these tablets of clay. Wind Cave is known as a barometric maze cave because changes in pressure cause wind to move in and out of it, and because it is, indeed, a maze. It has been called the most complex, three dimensional, rectilinear cave known. It is a giant fractal, created by a self-organizing process—containing many different structures shaped by many processes—all being driven into existence by the movement of energy. What message from the universe could I read here?

As we entered Wind Cave, I could already feel my chest tightening as a suffocating sense of claustrophobia descended on me. The extreme humidity in the cave made the air seem that much closer. I'd been in small, shallow caves, such as the ones in Copper Canyon where the Tarahumara live. But I'd never seen anything like this. Irrationally, I think I expected a neat hole like the Lincoln Tunnel with big rows of fluorescent lights. But as we left the mouth of the cave, we found ourselves descending through the bronchials and down into the lungs of the earth, a chaotic world of frenzied breakdown, man-sized holes that appeared without warning beneath our feet, everything shattered and thrown about in cascades of dark reddish rock. Each hole seemed to lead down to sharp rocks and more sharp rocks. Against this, we were equipped with helmets, headlamps, gloves, heavy knee and elbow pads, sturdy boots, and rough clothing as we plunged ahead, deeper and deeper into the labyrinth.

You have to see the cave to appreciate the seeming chaos of riven rock so complex that it made me dizzy just trying to take it all in. It was impossible. At first, I tried to look everywhere at once, to create a mental model, a map of my environment, but it just made my claustrophobia worse. Then, gradually, I began to focus on details. It took me a full hour to get the sense of how to move through the cave, not so much physically as emotionally. For one thing, the beam of my own headlamp was all the light I had, and if I swung it around, it was like the climactic scene in a horror movie. Monster shapes charged at me, and creeping red rock fingers reached out to grab me. There was no floor, so we were always on top of one rock or another, teetering, balancing, swept along, never quite still. As I looked down, I could see that there was nothing but the continuous descent to blackness.

At last we found a rock big enough to sit on and stopped to rest. Horrocks explained to me that the cave was, in effect, infinite. Since the first recorded explorations in the 1890s, Wind Cave has grown from a few miles to more than 117. The idea that a cave can grow might seem odd, but that growth represents the size of what's been surveyed. As the techniques for surveying and measuring the cave become more sophisticated, the cave grows in another way as well. Here's how it works. If you measure a coastline, for example, using a photograph taken from the space station, you might use a measuring device that is 10 miles in effective length. That measurement will give you one number for the length. If you then measure it on the ground using a surveyor's transit that is accurate to a mile, you'll come up with a new number for the length of the coastline. That number will be larger, because you've taken more detail into account. If you walk around measuring the coast with a yardstick, you'll get a number that is larger still. The more refined your measurement is, the longer the coastline appears to be. Theoretically, the coast is infinite, even while the area of the land it defines remains fixed.

Wind Cave can be thought of as a coastline in three dimensions. The length keeps getting longer within a fixed volume of land. No one can yet say what that exact volume is, since people like Horrocks keep finding new passageways. They know this much: it's big. Wind Cave breathes, as its name implies. Winds of 75 miles an hour have been measured at the entrance. When it breathes, it can inhale or exhale a million cubic feet of air an hour. Based on that airflow, Herb Conn, a legend in South Dakota for his rock climbing and an expert on its caves, estimated that the volume of Wind Cave is two billion cubic feet. Whatever processes are taking place here, they must be producing entropy, or else the universe would not have done all this work of excavation.

We moved on and on, climbing up huge boulders to a passageway so narrow that I had to crawl on my belly, pushing my backpack ahead of me as I dragged myself along on my elbows. We were about 300 feet underground, and the thought that I might become wedged in there filled me with dread. I couldn't help remembering the story of a young man named Thomas Edward Benning. One day, he told his wife that he was going caving. Caving can catch some people like a disease of the mind and obsess them like an addiction. The impulse to explore fractal structures probably originated deep in our past, and once a person who is susceptible to this passion begins exploring those labyrinths, it becomes impossible to stop. Benning entered Black Cave in Arizona some time in April 1995. His wife never heard from him again.

Almost two years later, in late 1996, three explorers entered the same cave. They had been moving deeper and deeper into the cave for about an hour when they found a book of matches. All the matches had been burned. The matchbook cover had been burned, too. Because of the complexity of this type of cave, normally only very experienced cavers go there. It's unusual to find anything man-made in a cave, because serious cavers do not leave their trash. Yet a few feet farther along, they found a burned-out miner's lamp. Curious now, they began exploring and found the insole from a shoe, a trash bag, strips of cloth torn from someone's clothes, a set of keys.

They were in a tunnel off to the side of the main line they'd been following. Puzzled, they continued on their original route. When they were ready to leave, they decided to look again at the side passageway. It was bigger than the one I was crawling through as I was having these thoughts. It was about three feet wide and 18 inches high. As they crawled through, the explorers in Black Cave found a shoe, then a sock, and then a scrap of fur wedged into a crack in the floor of the passageway. When one of the cavers pulled on the scrap of fur, however, it turned out to be the top of Benning's head. He'd been mummified.

Cavers love secrets. Benning may have thought that he'd discovered Black Cave. But local cavers had been mapping it and keeping the map secret. It's something in the nature of cavers. It's something in the nature of people. We begin making our complex mental maps at around the age of six. Between then and puberty, it is common for children to become fascinated with shortcuts and interesting routes and to trade them with friends and keep them secret from others. This occurs as the hippocampus, a structure in the brain that makes mental maps and serves to create new memories, begins to mature. I imagine cults of people doing this tens of thousands of years ago or more. Some of them left their drawings behind. The paleoanthropologist Bernard Campbell writes that "The most spectacular wall art is confined to true caves: deep underground fissures with long galleries and passages. These caves have their own subterranean pools, rivers, and festoons of stalactites and stalagmites. They are dark, mysterious, and very cold; they could be entered only by people holding stone lamps or torches."

Acting out this ancient and perhaps innate drive, then, Benning had gone down the side passageway, which was concealed from the main route by a boulder and which opened onto an unexplored region. His light failed. One by one, he lit his matches, trying to find his way. He used his last match to light the matchbook, which he dropped when it burned his fingers, leaving the first evidence that the explorers found. Then he began leaving a trail of his personal effects so that he could follow them back out—or so someone could follow them in—keys, insole, trash bag, strips of clothing . . . Desperate, cold, lost, alone, dehydrated, he crawled at last down into a crevice, perhaps seeking a way out, and was likely too weak to crawl out. Caves are really bad places to make mistakes.

So that is what I was thinking about as I elbowed my way along that passageway 300 feet below the ground with my heart hammering in my chest. I tried not to think of what Benning must have felt as the full realization that he was lost and would die descended upon him. I kept telling myself: Horrocks loves this cave. Caves are his life. He works for the government. He doesn't want me to die. Dead visitors equal bad publicity. Horrocks is not going to leave me. I'm not going to end up as a mummified piece of fur. (Well, I am, I suppose, but not yet.)

At last we emerged into a huge room, whose size was made difficult to comprehend by its complexity. As I shone my light down the side of the boulder on which I sat, it fell away through the branching bronchials of countless convoluted passageways—"leads," Horrocks called them—and I comprehended in the most immediate and physical way the expression "the bowels of the earth."

"Where does it go?" I wondered out loud.

Horrocks pointed out that my question was the first step in being seduced by the cave. You begin by wondering where something goes. Then you follow it. You vanish into the earth. We follow these leads as we follow ideas, complex sequences of images that become our explorations, our actions, our tools, our language, and ultimately our lives.

Horrocks coaxed me off my boulder, where I had perched in the hope that I wouldn't have to negotiate any more 18-inch seams. "I want to show you something," he said, hopping nimbly from rock to rock over crevasses that descended to the iron armature of the earth. I had already felt with my hands and cheek and forearms what this material was. It was sharp, like glass in places. It could cut, shear, and tear flesh. I couldn't imagine moving across it the way Horrocks did. Then again, like Pedro on the cliffs of Copper Canyon, he'd been doing it since he was a child. I thought, wow, he must have a huge hippocampus.

After a perilous traverse, I clawed my way to his side. I found him lying on his belly, pointing his headlamp into a small grotto. I lay down carefully beside him. It was cold enough that I could see my breath.

"Don't breathe too hard," he said. "It's incredibly delicate."

There, beneath an overhang, I saw what appeared to be a sea anemone, a silvery creature with thousands of sparkling hair-like feelers standing out several inches from its white dome-like body. Although the air was still, I had the illusion that the anemone was waving its tentacles in the sea of near-100-percent humidity in which we breathed. I was seeing the design of life, only it wasn't alive. It was the complex pattern, the order produced by some source of energy (maybe many sources) that had been spreading out here, struggling, perhaps for millions of years. Order isn't free. Something paid for this small treasure.

"What is it?" I asked.

"Those are gypsum needles," he told me. "Calcium sulfate. The same stuff that Sheetrock is made of." My mental model insisted that it was alive, but the creature was a stone.

Wind Cave is famous for formations known as boxwork—complex, interconnected blades of calcite that stick out from the walls in a honeycomb pattern. They formed after water dissolved limestone away from gypsum, which had been deposited in cracks in the earth. The boxwork was exposed when the water receded about 50 million years ago.

As Horrocks and I moved onward into the depths, we paused here and there to examine formations that seemed to my eye to be clear mirrors of the living world: boxwork bears a striking resemblance to the microscopic structure of certain animal tissues. In fact, a trip through animal tissue with an electron microscope will in many places eerily resemble a trip through Wind Cave. In caves like this you can find needles, beards of long hair, spiders, flowers, starbursts, trees, cells, and even a material like cottage cheese formed out of hydromagnesite, which sometimes expands into great jellyfish balloons. There are helictite bushes that looked like twisted trees, some of them six feet tall, growing from the floor. They grow against the force of gravity by capillary action. It seemed that we were getting a glimpse of the cosmic cheat sheet for the great plan of life. The second law at work here had produced stillborn children like the gypsum anemone. And when conditions were right, when the right materials were at hand, it would make living, reproducing creatures as well. But I gradually came to understand that the point wasn't that this looked like life. It was that life looked like this. And life looked like these structures because life formed according to the constraints of geometry and the laws of physics. Living or dead, these shapes were good for moving energy under these particular conditions and given the chemicals that were available.

There had been an entropy debt—or many debts—and building these structures had helped somehow to pay it off. Was the entropy debt chemical, thermal, or something else? What I saw around me probably represented many different processes, but they all had a common theme: when energy spreads out, creating organized structures, it allows a local decrease in entropy in exchange for a larger increase of entropy outside the structure. These organized structures occur at all scales, from galaxies down to the molecular components of cells and even the vortex of force that binds the quarks at the finest level of structure in an atomic nucleus. And they won't occur unless the universe has a use for them. What is that use? What sort of business is the universe so urgently carrying out? It's moving energy from a high concentration to a lower one. That's all we know. We don't know why.

As Horrocks showed me over the next several hours, there was actual life down here, too—an entire ecosystem of it deep in the earth, where the resources for life seemed not to exist at all. At the bottom of the food chain is a tiny mite that lives through chemosynthesis—that is, it eats minerals, just one more exception to what we once thought was a law: that all life depends on sunlight. From that mite, the ecosystem of Wind Cave moves up to a sort of hopping insect and gradually all the way up to bats and wood rats that navigate in total darkness by marking their trails with urine.

Horrocks and I had been exploring the cave for six or seven hours, while a team from the National Geographic Society was taking photographs elsewhere in the maze. When we decided we'd had enough for one day, Horrocks said we had to go find the others and let them know that we were headed up. On an impulse, I told him to go on without me, that I'd like to be alone. He looked at me, startled, and hesitated for a long time before answering. Perhaps he was thinking, as I was, about Rachael Cox, who was on an expedition in Wind Cave led by the National Outdoor Leadership School in the fall of 1989, when she wandered away from her group and lost her way. The intense search that ensued took several days, and Rachael was all but given up for dead. Following a gut feeling, some of the searchers went back to an area that had already been covered and heard Rachael's voice calling softly. She had somehow navigated to a lower level in the cave that no one even knew existed, and the rescuers had to excavate through the rock to reach her. It's very easy to lose your way in a maze cave and extremely hard to be found again.

"Look," Horrocks said at last, "if I leave you, you don't move, okay?"

"I won't," I said.

"I mean, you don't move at all. Don't go anywhere, do you understand me? Just sit where you are."

"Okay."

He studied me for a long time and then added, "Because if you go anywhere at all, I won't be able to find you." It was the first time since we'd entered the cave that I'd seen him display any concern at all. But I understood. I could see him asking himself, Is this guy a flake or can I trust him?

Then he scrambled away, and I watched as the beam of his headlamp bounced around and faded away.

With twenty stories of rock shutting out all earthly noise, a silence such as I had never heard before, not even in the Arctic, descended upon me. And I thought, if the cave swallows Horrocks, if he sprains his ankle or hits his head, I have no idea how to get out of here.

So I turned out my light.

Then the cave gripped me. There was nothing at all to tell me that I even existed except the hammering of my heart, my shallow breathing. Maybe this is what it's like to be dead, I thought. No, there was still much here to observe. There was still smell and touch, and there was the evident presence of my interior world, including my physical body with all its myriad signals and the vast landscape of memory and experience tracing its fractal way back to my childhood and, for all I knew, beyond that, too. I remembered my father showing me endoplasmic reticulum and mitochondria inside of cells and understood that, like this cave, a living being is an infinite landscape bounded by a finite volume. In our complexity, we mirror the universe itself. We are a bounded infinity, both bigger and smaller than all we see.

While waiting for Horrocks, I had time to think over what I'd seen that day. All this complexity had formed as energy tried to move from place to place. The same process caused the vortex in a bottle that I first saw as a child in a medical school laboratory. One of the messages here concerned the origin of life. New discoveries that had been accumulating since at least the 1970s would shed light on Murray Gell-Mann's statement to me.

I'd had the good fortune to meet Robert Ballard, the man who became famous for locating the wreck of the *Titanic*. But he wasn't a treasure hunter. He was a geologist working in naval intelligence, and he had spent much of his career studying plate tectonics. The molten rock that's beneath our feet hardens into the crust of the earth at the surface as it tries to release its heat into outer space. The crust is not one continuous surface; it's made of many pieces, like the cracked shell of an egg. The pieces are called tectonic plates from a Greek word meaning "builder." Most of the surface of the earth is under water, and the continents are simply the high spots on those plates that happen to protrude above the waterline. In certain places, lava pushes up through the cracks between the plates to create new material to form the surface of the earth. The new rock pushes the old rock away, shoving it along, and where it is in contact with another plate, its edge is shoved up and over or underneath, down into the molten rock, where it is melted and circulated again. Convection currents in the molten rock beneath help move the plates along. Gravity pushes down. The crust of the earth is not a uniform thickness. In places where it's thin, magma may pool and punch through.

Ballard had the idea of going to the ocean floor to observe the rift where two plates meet, and trying to determine what processes were taking place down there. He'd been assigned to Woods Hole Oceanographic Institute in 1967, and it just so happened that Woods Hole owned the most sophisticated manned submersible vehicle in the world, *Alvin*. In 1912, Alfred Wegener published a theory of continental drift based on the observation that if you look at a map of the land on one side of the Atlantic, it seems to match the shape of land on the other side. He believed that the two had once been one piece of land that had separated along a rift and slowly drifted apart. In the 1970s, there was renewed interest in the idea, and Ballard went to look for himself in an area of the Appalachian mountain range that lies underwater in the Gulf of Maine. During 1971, his dives in *Alvin* showed how the continents of North America and Africa had been pushed apart, riding on their respective plates, to create the Atlantic Ocean.

On the strength of that accomplishment, in 1973 Ballard was invited to join the largest oceanographic expedition ever undertaken, the French-American Mid-Ocean Undersea Study (the acronym was not lost on him). Inside *Alvin*, Ballard and his colleagues dove to 9,000 feet. It was the first time anyone had seen the edge of the North American continent, a 1,000-foot wall of sheer rock descending vertically to the vast central rift valley, which was scattered with undersea volcanoes. Across the valley was the Mid-Atlantic Ridge and, beyond that, the coast of Africa.

In early 1977, Ballard dove *Alvin* to a depth of 12,000 feet off the coast of South America. There he saw what he described to me as "one of the deepest and most spectacular submerged terrains on the planet." At that depth, there was no light except what *Alvin* provided. The pressure was 5,500 pounds per square inch. (Normal air pressure is 14 psi.) If the pressure vessel failed, there would be nothing left of the men and no one would ever know what had happened to them. Ballard described staring at the hex nuts that held *Alvin* together. He said it was "like gazing into the maw of a loaded cannon." He described his mouth going dry whenever the hull came in contact with rock.

The team of researchers believed that they must certainly have traveled beyond the limits of life to a place where nothing could survive. This, then, was the hard-rock earth of old, the baseline from which life would have had to wrench itself free. Or be seeded by life from outer space. Or . . . There were many theories, but the underlying assumption back then was that the universe is a cold, dead place, and life an exceptionally unlikely coincidence. So surely, by that way of thinking, nothing could live here beneath the ocean, where there was no sunlight, no oxygen, and extreme conditions.

As they settled to the bottom, they began to see something no one had anticipated: chimneys pouring out black smoke amid cottages and castles of volcanic rock, as if a medieval town had come to rest here on the ocean floor, all shimmering in heat waves emanating from the seabed. While there had been hints that such things might exist, no one had envisioned what they might look like or imagined their profligate complexity. These were the hydrothermal vents where lava bled across a huge temperature difference between the core of the earth, at between 9,000 and 12,000 degrees (4,982–6,649 C), and the icy ocean waters.

But there was more. Here, where the crust of the earth was being created, animals—creatures no one had ever seen or even suspected before—were living in the mineral-rich environment at temperatures of up to several hundred degrees. This superheated water by itself would have been enough to kill all known life-forms. The chemicals spewing out of the volcano would have been deadly, too. Yet here were creatures adapted to it—drunk with it and dancing to its music: giant clams with red meat interiors, blind crabs, and great tube worms, all based on a foundation of microbes that lived on hydrogen and sulfur. Ballard had taken a giant step back toward the moment of our own creation. He'd upended our model of things. Striving for the bottom, he had reached the top. He'd found the smokestacks on the roof of the world, one of the great factories in the supply chain that creates us all. But what was driving this process?

The implications of this find were astonishing. Here was a kind of life that took its energy not from the sun but from something deep inside the earth, where the heat of radioactive decay melted rock and churned up the world from below. It had begun to seem that life was simply a natural outcome of the laws and processes of our universe. There had been no unlikely accident. Life required no designer.

Over time, new discoveries gradually pushed the age of the earliest life on earth back so far that it appeared that life had formed the moment it was even remotely possible for the chemistry to take place. But at that time, something less than four billion years ago, the earth was being bombarded by meteorites and comets with cataclysmic regularity. How could anything live in such an environment? There was now a tantalizing answer: it could live deep beneath the surface of the sea, where things were relatively quiet.

One of the largest impacts—by an object believed to be the size of Mars—struck the earth and sent a debris cloud into orbit around the planet. The debris was eventually drawn together by gravity and became the moon. As the moon grew more dense and solid, its gravitational field served as a kind of shield to deflect larger objects, and things began to quiet down on earth. The first evidence of life that we know of dates from about 3.8 billion years ago.

Christian de Duve, who won the 1974 Nobel Prize in medicine, put it this way: "The Universe was pregnant with life . . . we belong to a Universe of which life is a necessary component, not a freak manifestation." Many scientists now believe this. As the cosmologist Paul

Davies has said, "All life feeds off the entropy gap." Where there is a concentration of high energy, there is low entropy. The high energy must move to spread out, increasing entropy in the process. That's the gap Davies is talking about. And it's also what Murray meant when he said, "The earth is rotting." He meant that high-energy molecules are decaying, so to speak, to lower-energy states. And when he said, "life is the waste," he meant that the structures we call life are created by that process to facilitate it.

At a hydrothermal vent as well as beneath the sea floor, there are hydrogen molecules that contain high-energy electrons. They represent a low-entropy state relative to the surroundings, and life provides a way across that entropy gap, a way of paying off that entropy debt by combining hydrogen with carbon dioxide, a chemical reaction that increases entropy and lowers the energy level of those electrons. That reaction can't take place without life.

All of this points to a view of life that is unlike the one that was being taught when I was a kid working in my father's lab, the one that many people might still find familiar. That view, now outdated, held that life came from tide pools and from an extremely unlikely combination of inorganic chemicals, encouraged perhaps by the energy supplied by lightning. This view was the result of an experiment done at the University of Chicago in 1953 by Stanley L. Miller and Harold C. Urey, which shook the scientific world at the time. But the universe doesn't like to do what's unlikely. It likes the most probable. This new view of life, then, begins to suggest some ways of answering the question that stymied my father when I put him on the spot in that diner long ago. Life is not strange or miraculous. It is simply what happens. If there is a sufficient entropy debt and the right chemicals, life will organize itself to move that energy along.

For many decades, fishermen had reported an odd material that they occasionally found in their nets. It was like dry ice, white, frozen, smoking—but, in this case, stinking as well. It smelled like a barnyard or like rotten eggs. In the 1960s, the same material was found in permafrost in Siberia. When scientists examined it, they found that it was methane hydrate, a crystal latticework of ice with methane trapped in it. Under high pressure and at low temperatures, water molecules form a structure like a cage around methane molecules, and those structures create the lattice that looks like ice. Magic ice. Ice that burns if you put a match to it. (Those cages, being ordered structures, represent an entropy debt, and won't form at ordinary temperatures and pressures.)

In the 1980s, scientists began to find methane and methane hydrate everywhere they drilled into the seabed. A map drawn by the Naval Research Lab shows methane hydrate fairly evenly distributed across the globe, though it shows only what has actually been discovered. (The earth is big, and research ships are few.) On one expedition in 1996, the scientists dropped a scoop half a mile down to the ocean floor off the coast of Oregon and pulled up a bucket containing a cubic meter of what looked like mud. When they dumped it on deck, it was bubbling and seething and smelled of rotten eggs. At the center of the mass was a core of methane hydrate. And what they mistook for mud turned out to be living microbes called archaea*—life, once again, where they thought none should exist. Numerous teams have gone out again and again, drilling as much as half a mile beneath the ocean floor, and no matter where they look, no matter how far down they go, they find the same thing: life.

These archaea (known as autotrophs—self-feeders—because they require no organic molecules to live) create themselves—out of thin air, in a manner of speaking. One type makes methane by joining hydrogen and carbon dioxide. Barnyards smell like methane because farm animals produce that gas in their intestines, as do people. We

*There are three main classifications of microbial creatures: bacteria, eucarya, and archaea.

have the direct relatives of those deep-sea creatures living in our guts today. Through billions of years we carried them, in a relay race of evolving creatures, up from the hydrothermal vents. Another type of archaea feeds on that methane and produces hydrogen sulfide, which smells like rotten eggs. The first organism gathers around the second, making what looks like—and is a precursor of—a shell. This is an example of how early cooperation among microbes could have eventually led to more complex organisms. Extremely small creatures like these, numbering almost a billion per ounce of mud, could have evolved into the shelled creatures that died in the sea and settled to become the limestone that acidic waters bored out and sculpted to form Wind Cave, where I now sat in total darkness, contemplating what Murray Gell-Mann had told me.

In the years since Ballard went down in *Alvin* to discover the vast communities living on hydrothermal vents, other scientists have traveled to places on the sea floor where methane seeps out. They have found similar communities, including the same types of tube worms and giant clams that Ballard saw. These communities have now been discovered across the globe. In some places, the scientists have seen long chimneys created by those shell-making microbes, the shell being the supporting structure of carbonate at the center, the outer layer being a community of microbes aggregated into something that looks and feels like flesh.

Some estimates now say that as much as 30 percent of all life on this planet is deep in the earth's crust in the form of archaea, some of which may be millions of years old. At the greatest depth, the methane producers take their energy not from the sun but from the entropy debt that is represented by high-energy electrons in hydrogen. There is no other way that we know of for the universe to move that chemical energy but to create these creatures. It seems that life first formed there, probably in the earliest time of the earth's cooling (within, say, about two million years, a short time in cosmological terms). Life as we know it—oxygen-breathing, light-seeking life—

is fairly recent. It took another two billion years to evolve. But the fact that life arises everywhere it can arise is becoming hard to deny. Every time someone has proclaimed that there is an environment on earth where life can't exist—a place that's too acidic, too cold, too hot, too dry, where the pressure is too great or there's no sunlight or oxygen—life has been found there. Life is found between the grains of mineral that make up rocks. Life is found clinging to the cores of nuclear reactors, where it would seem that the radiation would kill anything. *Deinococcus radiodurans* continues to live with 1,500 times the radiation that would kill humans and 100 times the dose that would kill the notoriously tough cockroach. One species of microbe lives in sulphuric acid. This may seem like a bit of a paradox, because although an extinction event can temporarily kill off most living things in a particular area, life will usually come back. But in human terms, that might not matter, since it could take thousands or even millions of years to do so.

I had lost track of time, sitting in the pitch dark of Wind Cave. My contemplations had led me by small steps to a better understanding of what Murray had told me. His statement now seemed not only to make sense but to be beautifully compact as well. Yet to see its elegance required that I know a lot of other details as well. His was a code, rather like the cryptic messages from the universe itself. To say that a man made that statement does not seem like such a big mystery. But to say that the universe made a man who made that statement seemed to me a lovely thing to contemplate: "The earth is rotting. And life is the waste."

Gradually a sound began to intrude on my silence. I heard a scraping noise. Was it wood rats? Bats? Something was approaching. Then a dim and flickering light began to play upon the rocks, and my vision returned by increments. As the fabulous red rocks came into being once more, I saw the light grow until the bouncing beam of Horrocks's headlamp lurched into view. I heard him clambering toward me with what seemed like increasing urgency. I realized that he must be seeing this dark room and thinking he'd lost me. He'd be thinking, I lost a reporter. Damn. That's going to look bad. And now I'll have to look for him all night.

I reached up and flicked on my own headlamp. I saw the look of irritated relief on his face. He exhaled noisily. "Let's go," he said.

Eleven

The 10,000-Watt Lightbulb

As I said in the previous chapter, the life-forms on earth, including us, reflect the nature of the universe and obey its rules. All animals and plants behave according to natural laws, as do the inanimate things of the earth and sky. Storms form and drop their moisture and rivers flow downstream, while mountains form and crumble, all by the same overarching principle (the movement and transformation of energy and the production of entropy). In the middle of all this, we humans stand in the strange position of being the only animal to think that we're not a part of it. At least we behave as if, somehow, deep down, we believe that we are above those rules, that we can step away and do something else by some other set of rules, which we ourselves will define, much the way a casino will design its games to give the house an edge. We are operating in a vacation state of mind and may forget that the laws of nature still apply.

It seems likely that life arose in the ocean. For three billion years, archaea worked away on the puzzling problem of energy transformation. Then, in a dazzling series of brilliant transformations, life migrated across time to find us in the person of Lucy and *Homo erectus* and eventually in the people who drew the subtle and beautiful images of horse and red bear and mammoth and rhinoceros and even a nude Venus on the walls of the Chauvet-Pont-d'Arc cave some 32,000 years ago in the place we now call France. The genius of this big-brained hominid was suddenly evident everywhere, as invention spawned invention.

Loren Eiseley described our progress on this planet as "a climb up the heat ladder." One of the most significant transformations ever wrought by an animal occurred when early people put fire to meat and crossed over into an ever-expanding source of high-energy protein, which was otherwise available only in awkward bits. Fire made meat clean and easily digestible. Fire allowed us to set the plain aflame and drive entire herds of elephants to their death. Fire made for plenty and made it possible for us to move energy around and produce entropy faster than any animal had ever done. We burned through the energy of the food we ate and the energy in the world around us, as we set fires to clear the land and organized it into plots suitable for corn and cows. From there we climbed the ladder all the way up through internal combustion, electrical power plants, jet engines, and air-conditioning. Seen from the space station, the dark side of the earth is aflame with light from the energy we're transforming at a furious pace, producing even more light than the rolling thunderheads that keep their nightly pace around the globe. But lighting represents just 8.8 percent of the electricity we use in the U.S. We use 35.5 percent of our electricity just to cool things off—air-conditioning, refrigerators, and freezers. We use another 10.1 percent for heating (most homes are heated by natural gas), and another 9.1 percent to heat water.* Entropy is all about heat. And we are prodigious producers of entropy in keeping with the second law of thermodynamics.

If we arose from nature and from natural law, then doesn't this

*Figures on electrical consumption come from the Web site of the U.S. government Energy Information Administration at http://www.eia.doe.gov/.

mean that everything we do is natural? This is a deep puzzle. We can trace ourselves as a natural phenomenon back to a time near the end of the age of reptiles, when the first flowering plants appeared, as if by a miraculous coincidence, just when mammals and birds would have needed them. Those flowers, like many emergent phenomena, didn't proceed by the gradualism we usually (perhaps mistakenly) associate with natural processes. They exploded on the scene and were suddenly everywhere. And just in the nick of time, they supplied an abundant source of high-energy food for those new creatures who, unlike their predecessors, were warm-blooded and needed much more energy to run their hot metabolisms and their ever-expanding brains. (Our brains, which make up only 2 percent of our body weight, use a fifth of all our energy.) Warm-blooded animals, such as birds and possibly even dinosaurs, may have come before the angiosperms, but with the appearance of those plants, with their concentrated source of energy locked inside of fruits and seeds, the profligate spread of birds and mammals was ensured. That event is known as the Cretaceous explosion. It was as if the entropy-producing system that was life had shifted gears all at once, suddenly becoming capable of processing energy that much faster. This has all the earmarks of the type of change associated with a self-organizing system like the vortex, which transforms itself into a structure for allowing a greater flow of energy.

But who was running the show? Were the animals using the flowering plants, or were the flowering plants using the animals—because it was the angiosperms that devised the countless ways of transporting their seeds by using mammals and birds, from the burr that catches in fur to the seeds in fruit, which will pass unharmed through the body of a bird to fall to the ground conveniently encased in fertilizer. Eiseley took the view that flowering plants, in their great wisdom, built birds and mammals for their own purposes. By that view, they built us, too.

Starting 100 million years or so ago, everything that evolved seemed to serve the flowering plants in some way. And as the wise flowers moved their energy through these creations, into buffalo and saber-toothed tiger, they were doing the remarkable work of the second law. In the old days, the age of reptiles, things had moved more slowly. Trees and reptiles moved slowly and processed energy slowly. Suddenly, with the appearance of birds and mammals, we were ripping through unprecedented amounts of energy at ever-accelerating rates of flow. When man came on the scene with his tools and fire, that flow rate took another leap. At each historic change in evolution, the flow of energy, the production of entropy, seems to have increased once more. Then, too, these systems sometimes collapse in extinction events. There appears to be a continuous striving in nature, but nothing grows to the sky. If you pulled back and took a very long view of this, with living creatures piling one on top of another, then collapsing all at once, only to begin again, you might be reminded of the sand pile with its avalanches at all scales. Indeed, extinction events vary with the same sort of mathematical regularity.

Viewed in this way, as a self-organizing system for processing energy, people seem to be in some odd sense exactly what we like to say we are: the pinnacle of evolution—not because we are smarter or have a soul or do open-heart surgery, but because we are able to produce entropy so well. What species before us could reduce a beautiful ancient city to rubble in a matter of seconds, as we did with Hiroshima? We're taking all the energy we can find stored in the bowels of the earth from all previous plant and animal life and doing what it could not do on its own. We're burning it up at a rate of 80 million barrels of oil a day, to say nothing of coal, natural gas, and other forms of carbon. Where once blue skies greeted me in the morning, now I see the tic-tac-toe of contrails, as jetliners carry carbon from the bowels of the earth and spew it out in the most ecologically sensitive place: the stratosphere. The evolutionary biologist John Whitfield wrote, "A woman in the United States uses as much energy as a 30,000-kilogram primate would, if such a behemoth existed." Geoffrey West, president of the Santa Fe Institute, put it this way: "Humans require about 100 watts of power. We need the same amount of energy as a lightbulb. This shows how extraordinarily efficient we are. So if we turn off one of the lightbulbs that we leave on all the time, that energy could be used to support someone starving in Africa, for example." But although we each need only 100 watts to stay alive, in the United States each of us burns energy at 100 times the rate of our natural metabolism. It's as if each of us has a 10,000-watt lightbulb that burns day and night, 365 days a year.

We are masters of transforming energy from a useful state to a useless state, which is what the second law says must be done. When the universe has a hot spot, a reservoir of energy that won't spread out readily, it invents new ways to accomplish that task, as we've seen.

Problem: A natural reservoir of energy exists in all that hydrogen locked up in oil and coal and gas in the earth. Bacteria can feed off some of it, but most of those hydrocarbons seem destined to remain locked up indefinitely. It's a big entropy debt, a big marker owed to the casino.

Solution: Invent people.

We can now view a subdivision of homes, each burning its 10,000-watt lightbulb, like a rain forest—a natural structure optimized for degrading energy.

War is another very good man-made way to produce entropy. If you think of money as a proxy for the accumulation of human work, which represents our human effort in creating order, then spending that money to produce nothing but heat is certainly an effective way to increase entropy. And war is a way to turn money into heat at a rate that no other human activity can equal.

From this point of view, we appear to be doing exactly what we're supposed to be doing. But if we had any sense, we might be doing less of it or doing it in a different way. The big question facing us is whether we can go against the house rules and stop producing so much entropy. Are we capable of becoming the first species in history to defy the commands of the second law of thermodynamics?

The main difficulties are no longer technical ones. The difficulties come from our own nature. And hopeful signs are there that we can change. We have the know-how to make those changes, and, as Amory Lovins put it, "No changes in lifestyle are required." In short, we have the opportunity to truly be the smartest creatures on earth and to do the smartest things, too.

The need for change is urgent. We must not only face the problems squarely, we must internalize them as part of our model of the world. That's hard to do, because we receive so many signals that tell us everything's all right, that our strategy is good. At least in the United States and Europe and a number of other countries of great advantage, we are awash in sensory signals that tell us how well we're doing. Here in northern Illinois where I live, the ground is covered with snow, the sky is overcast, and there's a freezing fog settling on the glistening tree branches. But I sit in shirtsleeves in perfect comfort, delighting in the smell of frying potatoes wafting from the kitchen. Soon I will eat that food and be even more effectively rewarded. And as I do, holiday lights blink and my furnace burns hydrocarbons just to make me happy.

We do what we're rewarded for doing. But as with everything else in our universe, this wonderful state of affairs comes with a cost, and that cost is hidden from the mechanism inside me that interprets these rewards to direct future behavior. That means that we must be more suspicious of our rewards and look more closely at what we've done to earn them.

I've already talked about some of the costs, such as heating up the earth. Commercial fishing has done so much damage to the life of the sea that, despite more efficient technologies for catching fish, the harvest falls each year. Each year trawlers dragging nets across the bottom of the ocean destroy ecosystems that amount to 150 times what logging removes in the same period on land. And that is to say nothing of the mercury and other chemicals that we dump into the sea. I live on one of the greatest freshwater seas in the world but can no longer eat the fish, which are now poisonous.

We're lucky to have the freshwater, though we may not be using it very wisely. It takes only one gallon of water a day per person, at most, to sustain life. In the United States, each of us uses about 75 gallons a day. About 80 percent of that water is used for three purposes: bathing, washing clothes, and flushing the toilet. How should we view that activity? Moving all that water around produces a lot of entropy, which appears to be what the universe wants. On the other hand, we can't use water at this rate indefinitely.

Similarly, when a company employs thousands of people to produce millions of products in a process that transports vast amounts of energy, and when those products are then thrown away to be replaced by similar products, that creates more entropy than would be created if that company did not exist. Is that a good thing or not?

Although Orwell wrote this passage in reference to war, it is equally correct to say that our consumer culture is "a way of shattering to pieces, or pouring into the stratosphere, or sinking in the depths of the sea, materials which might otherwise be used to make the masses . . . too intelligent." Orwell saw the effects he described in his novel *1984* as a conspiracy. But no conspiracy is needed. We are just part of another self-organizing system doing precisely what we'd expect such a system to do. War shares many characteristics with such systems (like earthquakes and extinction events, the death tolls from wars follow a mathematical distribution known as a power law). As I watch these phenomena, I can't help but ask how our natural tendency could be otherwise. Are we misinformed? Greedy? Evil? Or is nature just subtly and secretly having her way with us? Can we choose what our behavior will be? Are we smart enough yet? If we understood that it's possible for us to stop wasting our resources and still have just as much fun as we've been having, then we might choose differently.

Let's look at the great organizing forces of our world: thermodynamics, quantum mechanics, gravity. But also evolution. Evolution has brought us to a place where we can be aware of—and perhaps choose—the way we affect the life of the planet and everything around us. For the first time in the history of life on earth, we have before us the possibility, the hope, of acting with greater deliberation in everything we do, from how we use our resources to how we choose to populate the earth. We now have the ability to know the rules.

We seem to have individual free will, in that we can choose trivial things in our lives, like whether or not we exercise or put money in the bank. But can we save ourselves from the fate of being agents of entropy? If rational people with good information still cannot make decisions that we can all see are necessary for our own survival, then how can we explain our behavior? We are part of a system whose baffling behavior is the product of numerous individual agents taking actions according to a simple set of rules. For people, we might state the rules this way:

- Acquire as much energy (or atoms or some other conserved quantity) as possible as cheaply as possible.
- Use it up as quickly as possible. (That is, increase entropy as much as possible.)
- Find some strategy to increase this process.

We can view this on a universal scale, a global scale, a national scale, or a personal scale. It is similar at all scales. On a personal scale, the nature of ape culture means that status counts for a great deal in achieving those three goals. Status gives access to more resources, and having more resources (energy, atoms) brings more opportunities to reproduce.

In the case of people who destroy their own environments, the end result is not that, in a few years, the place where we live will be uninhabitable; it is that in a few years, it will be uninhabitable by humans. Although it might take a few million years, other creatures will eventually live there. Viewed on a universal scale, humans are irrelevant to the process except to the extent that they occupy a position, temporarily, as agents of entropy. When the process of degrading energy is through with us, when we have served our purpose and are no longer useful to the flow of energy, the system will discard us, and another set of agents will take our place. The loss of humans in this local portion of the universe has no effect on the larger system. As Marcus Aurelius put it,

> The universal nature out of the universal substance, as if it were wax, now moulds a horse, and when it has broken this up, it uses the material for a tree, then for a man, then for something else; and each of these things subsists for a very short time. But it is no hardship for the vessel to be broken up, just as there was none in its being fastened together.

We may hear it said that life is fragile. Life is not fragile. Your life is fragile. My life is fragile. But life occurs everywhere that the right conditions and chemicals happen to be. The planet will do fine without me. But most of us would nevertheless like to safeguard our own lives and those of our loved ones. Our kind are hoping to live on this planet for a very long time. Can we use free will to contravene the second law of thermodynamics in order to leave something for our children? We know that we have the technical means to do it and even to make a profit doing it. But will we?

This is a deep philosophical question, for which no answer exists at the moment. What makes it a philosophical question as opposed to a political one or some other kind of question is this: it's actually cheaper to stop living the way we're living. We could actually improve corporate profits, make the economy grow faster, and enjoy a better standard of living if we made the decision not to squander our resources. For example, for the price of a few months of war (90 billion dollars) we could retool the transportation system (cars, trucks, and planes) and triple its efficiency. By spending the same amount on biofuels for that newly efficient fleet, we could make 70 billion dolars a year in profit. To people concerned about the environment, it may come as a surprise to learn that the Pentagon may emerge as a leader in this area, because they are the largest slave to oil. Much of their activity involves hauling around cumbersome and vulnerable loads of carbon fuels in order to accomplish their missions. About two-thirds of everything the military has to move is fuel.

Throughout history, philosophers have wanted to know about the nature of free will and reason, and we are in a unique position to find out. The biggest question looming for the human race is whether we can actually use reason to change our behavior and to do what is necessary to survive, or if we will simply be carried along by a process that is already under way.

Twelve

The Inside–Outside Problem

In the beginning of the universe, there were vastly fewer choices for doing the work of the second law of thermodynamics. Energy spread out. There were few circumstances that could create what Eric Smith, of the Santa Fe Institute, calls "transport channels." While this term hasn't been widely used, I find it useful in visualizing the way many processes work to increase entropy. A transport channel is a structure that speeds the transport of energy as it moves to pay the entropy debt.

You'll recall that an entropy debt needs to be paid whenever there is a reservoir of high energy in a setting of lower energy, such as a hot cup of coffee in a cold room. The heat energy of the coffee cup has no special channel to move through, so it will simply dissipate by conduction. If you set a cold pan on the stove, it will heat up through this same chaotic process. But as we've seen, sometimes channels form that speed the transport of energy.

Lightning is a familiar transport channel.* It is formed when

*I'm leaning heavily on the work of Harold Morowitz and Eric Smith in these explanations, especially a 2006 paper (see Bibliography) and a lecture called "Inevitable Life" that Eric gave at the Santa Fe Institute.

there is a difference in charge between the earth and the air. When the difference is big enough, the air becomes ionized along a narrow path, like a tube, and electrons can easily flow down this tube into the earth, evening out the distribution of charge and increasing entropy through heat. The temperature of the air rises sharply along the channel of the lightning bolt. Since hot air expands into the surrounding colder air, the work of moving all that air is done. We experience that movement as sound: thunder. The difference in charge could eventually even out without lightning, but it would take much longer and would not be as complete.

Another common type of transport channel is formed in convective storms. The easiest example to visualize is a hurricane. It is a very clear structure, a well-defined vortex, driven by a difference in temperature between the ocean and the upper atmosphere. The storm creates a cell in the middle, around which the vortex turns, precisely like the vortex in a bottle. Both lightning and hurricanes exhibit an important feature: they create an inside and an outside—an organized structure, the channel, through which energy flows.

These phenomena are both referred to using the term "nonequilibrium," because for the structure to get started requires that two areas be out of equilibrium, such as a warm ocean beneath a cold layer of air. Somehow energy must have accumulated unevenly—that is, out of equilibrium. In the summer, when the earth tilts my home more directly toward the sun, thunderstorms and tornados break out in the area where I live. They do so because the sun is causing a rise in temperature in this area. That creates a local entropy debt, and the only way to pay it off is through those vortex structures we know as convective storms that move lots of energy around. While this process is perfectly natural, it is not always good for us.

Knowing what's going on can help us to avoid becoming coupled with those systems. I don't hike the ridgetops of mountains in the presence of transport channels, lest I become one. But I view the understanding of science as more than a strictly practical matter. There is a vast and secret world right beneath our feet, a great drama playing out all around us, and very few people appreciate it. Most of those who do are scientists and fewer still are good at explaining it to nontechnical people like me. I've been lucky enough to meet some of those few, and I believe that we should all have a glimpse into the theater where our lives play out to ensure that we're not just puppets but active players as well.

As should now be obvious, the vortex in a bottle is a transport channel. As I swirl the water, making it move in a circle, a structure spontaneously forms that has an inside and an outside. The vortex itself is the inside. The outside is everything else in the universe. The vortex is a walled structure like the hurricane. It has an "eye" through which the relatively high-pressure air in the room pushes up into the bottle, as gravity pulls the swirling water down in a spiral shape around it.

A living thing or an ecosystem can be viewed as a transport channel for energy, too, chemical energy in this case. As Harold Morowitz and Eric Smith wrote, an atom that can use photosynthesis is "a billion times more efficient as a transporter of energy" than nonliving matter. There are only two opportunities on earth for the type of transport channel we call life to arise. One opportunity comes from geothermal and geochemical energy. Those forms of energy create the deep-sea colonies that Ballard saw when he descended to the ocean floor in *Alvin*. Hydrogen won't react with carbon dioxide by itself, but the archaea that produce methane can make it happen. The other opportunity for the creation of a living transport channel comes from sunlight, which produces the more familiar life-forms we see around us.

The first type of life uses fission—splitting apart of atoms—for its energy source. That comes from the radioactive core of the earth, which gives off enough heat to melt rock. The other uses fusion, the binding together of atoms that takes place inside of stars like our sun. Without going into the details of the chemistry involved, both types of systems increase entropy by transporting electrons from a higher energy level to a lower one by way of what's known as the citric acid cycle. The archaea run this cycle one way, and we run it in the opposite direction, which we call the Krebs cycle. The archaea run the citric acid cycle to assemble the more complex biological molecules that make up their bodies. We run the process in reverse to digest the bodies of other living things. We eat dead things. Archaea eat rocks.

Either strategy works as a transport channel to move energy and increase entropy. As Eric put it, what kind of life you are depends on how you handle electrons. Reductive life, like archaea, works by using high-energy electrons to make organic molecules. That process lowers the energy of those electrons, which increases entropy. Life that depends on oxygen (like us) works by a more complex process, but it, too, involves the handling of electrons, and it actually embodies the processes of reductive life within it. Subsea archaea would be killed by oxygen, hence, as Eric said in one of his lectures, "Oxidative life is reductive life wrapped up in a space suit that's able to capture sunlight."

Whichever process a given organism uses, all life arises from the simple citric acid cycle, which involves just eleven molecules to facilitate the flow of energy. A key feature of that cycle is that, as the molecules go around, they get larger. At a certain point in the cycle, the largest molecule splits in two, and the cycle begins again. The process of moving energy (and specifically of moving it to a less energetic state) builds increasing structure and order as it goes around the cycle. While that structure represents a lowering of entropy locally, it serves the purpose of producing more entropy globally. As with the games in a casino, there is an inescapable cost that must be paid. There is a net increase in entropy.

This creates an inside–outside dichotomy. But it also unites the inside and outside through the flow of energy. Energy comes in to build the structure and flows out again as the products of metabolism. Those products are the currency with which we pay the entropy debt. As more and more complex molecules are built with that process, evolutionary selection can begin to take place. Molecules that are better at moving energy will tend to win.

Certain deep laws and principles unite everything in our world, including us. Our ability to understand our lives and behavior and our predicament in this world depend on our understanding those laws. As we go down and down through the layers of understanding, we eventually come to find ourselves up against the sometimes daunting subject matter of thermodynamics and quantum mechanics. Without the lens of mathematics to focus the picture more sharply, we are left with an incomplete understanding. But you can still play in the casino without a degree in statistics, and even without mathematics it's illuminating to recognize that these systems that process energy, which arise spontaneously, are at the heart of the effort of scientists to learn how life first arose on this planet. By this view, it would have had to arise in many other places in the universe as well.

Living things reflect the inside–outside dichotomy in the very delicate balance involving individuality. They have to maintain the self—the channel—as distinct from the rest of the world. At the same time, they have to take resources from the world and incorporate them into the self without harming the self. They have to transport energy from outside to sustain the self. As life has changed through evolution, those inside–outside problems have multiplied. For example, the subsea archaea simply take in high-energy chemicals and reduce them. To reproduce, they split in two.

Organisms that reproduce sexually face a much more knotty problem. They have to relinquish part of the self and blend with another individual. When you reach the level of animals like dogs and cats, the proposition can be downright risky. Sex can be a violent matter. The mechanisms of emotion tend to be generic, and larger animals, equipped with claws and fangs and antlers, can hurt each other when trying to exchange genetic information. Giving up what is necessary for reproduction while keeping the self intact—that is, maintaining as clear a boundary as possible between the inside and the outside—can be tricky. Indeed, sexual reproduction appears to be a strategy for keeping the self intact. By splitting the genetic material for the offspring, it makes the offspring a less appealing host to the parasites that each parent is carrying. The offspring has a head start in keeping the outside at bay.

This inside–outside strategy is also the aim of the immune system, which we could say is the key to defining what "inside" means to an organism like us. The immune system is there to keep the outside out. It keeps a close watch on what comes in and only allows outside materials to remain when they will help to build or fuel the inside self.

People form clans and tribal groups that have sturdy boundaries between inside and outside. This is the groupness effect discussed earlier. Those social structures have very strong immune systems that reject anything from outside the group. As we've seen, individuals from other groups may be killed. But the group needs genetic diversity to be healthy, so groups may have a strategy by which they raid other groups and kill everyone but the females of reproductive age. Other groups may have a more peaceful arrangement by which either males or females migrate to a new group. This is a risky but necessary step, and coming into a new group at the bottom of the status hierarchy is painful, as Viaje learned at the Milwaukee Zoo. But it admits just enough new genetic material to keep the inside healthy.

Going back to the idea that there were fewer choices at the beginning of the universe or of life, it would seem obvious that the processes of life down below the sea would all be automatic—that is, those processes wouldn't have to be associated with anything that we could call thinking and deciding. That's why those life-forms are crowded around the hydrothermal vents: that's where the energy and chemicals to create and sustain life are. If you put a microbe in a lower concentration of nutrient, it will move in the direction of a higher concentration. It will never stop and decide to go the other way just for the heck of it. It will never go on a diet. Automation is the rule. Any living thing that doesn't automatically do the right thing loses the evolutionary lottery.

Going forward in evolution, once again, we see more complexity, more choices—and more need of some organ of choosing. In a sense, the microbe is looking into the future by moving from a low concentration of food toward a higher concentration. It is predicting that conditions will be more favorable in the future that is represented by "over there." (You can't move through space without also moving through time.)

But much more sophisticated means of predicting the future evolved. The senses are essentially a time-shifting strategy. If you are a truffle, and I am a pig, you'll never see me coming. I'll just eat you (or dig you up for the French chef who has me on a leash). If I'm a rabbit, and you are a hawk, I can sense your position in space (using eyes, ears, nose) and calculate where you will be at a point that lies a few seconds into the future. This time-shifting represents an immense advantage. The rabbit can predict the future in precisely the way that is best for its own survival. But it now has many choices of how to react to that information—jump, crouch, go left, go right, backward, forward, down a rabbit hole, make noise, remain silent, empty bowels . . .

In fact, there are so many choices that the time-shifting device does no good if the rabbit takes too long to decide. There has to be some quick and dirty method for translating that slim edge of time into an immediate action, one that beats the flying speed of the hawk at least some of the time. The rabbit must not be too slow, or there would be no rabbits. It must not be too fast, or there would be no hawks. The two form part of a system of life. A characteristic of this system is that both the rabbit and the hawk must make mistakes. They must fail part of the time. Natural forces fine-tune the system and set the rate of failure precisely. When the system goes out of tune for whatever reason, one or another of the species (or both) will become extinct, as many species have done when people arrived on the scene. When people wish to do something like building a housing complex in the habitat of an endangered species, this is the system that they need to understand. This high degree of interconnectedness and the delicate tuning of the system make it fragile. That is the reason that it's important not to kill the little blue heron, the snowy egret, Indian cucumber, and numerous other seemingly irrelevant species.

The rabbit's mechanism of choosing relies on the underlying assumption that I've discussed: what has happened before will happen again, and what hasn't happened won't ever happen. The rabbit starts with an innate reflex—to duck whenever any shadowy form looms overhead—and then refines that through experience and through imitating other rabbits. This process creates mental models to identify things quickly and behavioral scripts to react to them automatically. An adult rabbit won't react to many harmless forms, only to hawks or other predators. I have watched this drama play out in my own backyard, where we have red-tailed hawks, peregrine falcons, giant barred owls, and plenty of rabbits.

That same primary heuristic is intact in us today. It operates alongside another system that conveyed a survival advantage: sequential thinking. Deliberate thought and logical, stepwise reasoning. This confers advantages in our ability to hunt cooperatively, to create the mental and physical steps to make complex tools, and to put together signs or words to communicate more effectively: language. But now we have two systems that are in competition with each other. The earlier system, which can be generally called the emotional system, has retained its dominance in many situations. In a perceived emergency (and often without one), most people tend to react without thinking by activating whatever seems to be the most appropriate behavioral script. Like Sten Molin in the cockpit of American Airlines flight 587, they may be wrong. Nevertheless, his was an intelligent mistake, one made through the activation of a generally reliable system for learning. To the organism that learned it, it made sense at the time. The system for learning is intelligent. The output of that system may not be in every circumstance. In the end it may turn out that we are like superfast hawks: we are using up all the rabbits. The difference is that we have the capacity to care that we're doing that. Whether we will do anything about it still remains an open question. To do something useful, we have to take two steps. First, we have to know. Then we have to care.

Thirteen

Why We Care

A lot of different strategies for survival are represented in the species that live today. Some of the older creatures on earth hardly move. Somewhere along the line, single-celled creatures attacked, ingested, or otherwise merged with one another and began cooperating. Some of the structures inside of the cells that I first saw as a child with my father's electron microscope are thought to be remnants of that struggle between inside and outside. Mitochondria, the power plant of the cell, were probably some kind of bacteria that got inside the cell and gradually evolved to be dependent on it. The meshing and merging and mutating of inside and outside proliferated wildly.

About 2.5 billion years ago, cells with a nucleus (an inside within the inside) evolved, and these so-called eukaryotes developed sexual reproduction, a new way of mixing genes that led to much diversity. A dividing cell reproduces a clone of itself, more or less, and only random mutations—mistakes—will provide the opportunity for change and diversity. Sex, in effect, supercharged evolution and thereby accelerated the transformations of energy that would increase entropy. Somewhere along in there (the dates are vague), complex soft-bodied animals, called metazoans, appeared in the ocean waters. Instead of being all one kind of cells, they had tissues differentiated for specific functions. Most had a digestive system of some sort. Those represented two radical strategic innovations: division of labor and mass production.

In the period around 500 million years ago, the metazoans began to grow mineral support systems, early skeletons. A really advanced life-form would have been a sponge or a coral. The plants were no more complex than algae. But trilobites also appeared, and they really managed to get around. Directed locomotion conferred another advantage.

Sometime over the next 100 million years, early fish began to appear. But mass extinctions also occurred. About a fourth of all marine families and 60 percent of the genera* failed and vanished around 439 million years ago.

New species evolved to fill the empty niche, as fish devoloped jaws and early plants spread over landmasses. Around 350 million years ago, sharks were prevalent, and the first creatures to breathe air—spiders, mites, and amphibians—came out on the land. Forests of primitive plants spread over the earth. And then another wave of extinction culled about a quarter of all families and put almost 60 percent of genera out of existence. But once again life flourished and surged forward into more complexity.

Around 300 million years ago, while coal was condensing out of sediments laid down in swamps, the earth became cooler and drier, causing glaciers to form. Cockroaches became one of the most suc-

**Genera* is the plural of the Latin word *genus*. Each living thing is identified according to these nesting categories: kingdom, phylum, class, order, family, genus, and species. For example, people are in the animal kingdom, phylum of cordates (we have a spine), class of mammals, order of primates, family of hominids (along with other apes), genus of *Homo*, and species of *sapiens*.

cessful species ever known. Trees that we might recognize as familiar today began to cover the earth, and early reptiles came into their own. By about 250 million years ago, the landmasses on earth formed a supercontinent known as Pangaea that extended almost from pole to pole. Dragonflies and beetles had appeared, and reptiles had become a dominant predatory life-form.

This is where the story begins to get interesting for us. Reptiles represented a new twist on the strategy for life. They could move around and eat things. Their brains were little more than dim lightbulbs on top of a spinal cord, and accordingly they were capable only of a rudimentary system of responses (compared to us) that allowed them to attack prey and to mate. They laid eggs, and when some reptiles hatched, their first job as babies was to get as far away from mom as possible, since she would eat them, too. But some had begun to care for their young in a rudimentary way. This was a new idea.

Then another extinction event, the largest in history, wiped out 99 percent of all life. Known as the Permian mass extinction, it happened about 250 million years ago. Although lots of theories have been put forth to explain mass extinctions, there doesn't have to be a cause (and this one was not caused by a big rock striking the earth). Just as in a stock market crash or an earthquake, such avalanche events may simply occur as part of the normal functioning of a self-organizing system.

Life did surge forward once again. But it is worth noting that on a human timescale, this really was a catastrophe. Douglas Erwin, a paleobiologist and director of the National Museum of Natural History at the Smithsonian Institution, pointed out that it took four million years before life began to come back, and in that time there were only about two dozen species alive in the oceans. In a lecture at the Santa Fe Institute, Erwin said that many species were still around for a while, "but we call them 'dead clay walking.' They were doomed but didn't know it yet." Why? Because key elements of their ecosystem were gone. They were hawks with no rabbits. It's useful to know about these prehistoric events; this sort of knowledge is the knowledge of the possible. When we build a shopping mall that causes some obscure and seemingly unimportant species to go extinct, we may be sealing our own fate as dead clay walking.

Some time between 250 and 200 million years ago, the first dinosaurs and early mammals appeared; and near the end of this period, known as the Triassic, another wave of extinction wiped out about three-fourths of the life on earth. At around the same time, that seam in the earth where Ballard first saw plate tectonics in action erupted. The explosion was so powerful that rocks from it can be found from Brazil to Spain, and in the eastern U.S. It split off Africa from North America and lava pushed up through the cracks and spread the plates to form the Atlantic Ocean.

By about 150 million years ago, dinosaurs were the dominant land animal, and another innovation evolved in the form of archaeopteryx, the flying dinosaur, predecessor of modern birds. Mammals existed at that time but were marginal niche dwellers compared to the dinosaurs. And then another wave of extinctions swept away the dinosaurs and many other species, leaving a roomy planet where mammals could evolve. At the same time, as mentioned, angiosperms, those flowering plants, appeared to catch the sunlight and pass high-energy molecules to us so that we mammals (and warm-blooded birds) could begin our dance with them, which is still under way.

That period, beginning about 65 million years ago, was marked by a proliferation of genera and species within our class, the mammals. This was the Cretaceous explosion mentioned earlier. A big part of our strategy for survival involved forming social groups and caring for our young and for others in our own group. Sometime within 10 million years of the extinction of the dinosaurs, our ancestors evolved into creatures who, instead of laying eggs, gave birth to live young and took care of them until they had developed relative independence. They nursed and licked and groomed them and withdrew them from harm. Being effective in those kinds of interactions required a bigger brain. In addition, a special layer of brain material developed, which has traditionally been called the limbic system. (Reptiles have some limbic tissues, but in mammals, the emotional brain developed into a rich complex of structures.) These new creatures developed emotional lives.

Emotions may have existed before, but mammals could now learn all sorts of new responses that would take them well beyond reflex behavior. With a new depth of memory, they could automate their responses to create complex behavioral scripts. They could learn a great deal more and store more intricate mental models. They became more adaptable and flexible. But the most remarkable gift the limbic system gave mammals was groupness, a new sort of inside–outside relationship. And one requirement of groupness is that, to cooperate in a group, you have to care. A rudimentary form of caring was born and perhaps a rudimentary form of love as well. An emotional system was not only an essential tool of the new strategy, it was the ultimate force in crossing the inside–outside barriers. It gave us the ability to read the internal state of another animal—and to change and be changed by it.

We're still evolving. You might say that we are somewhere between apes and angels now, and we have much of the ape in us still. That fact shapes some of our most important human activities. More than most people would like to admit, it shapes one of the most important activities of all: how we manage our relationships with other people, our groupness. Relationships have always been there in some form, but ours have evolved so far that they represent an important qualitative change.

All living things communicate in some fashion, even if it's just by way of chemical signals. As the brains of mammals grew larger, their language grew richer and more flexible. Birds have some of these traits, too, but mammals also have a completely revolutionary means of communicating. They have faces.

Facial expression is a universal form of communication among modern mammals. This silent, secret system is made possible in part by the fact that facial muscles are the only muscles in the body that attach directly to the skin, which allows for deep subtleties of expression. Body language is nearly as expressive, with its panoply of attitudes to display emotion. These are just two of many rich channels of communication that allow one creature to read another's internal state and to alter it and be altered by it.

Everyone knows that you can say a great deal without speaking a word. Potent messages can be sent by facial expression and body language, as well as by tone of voice and even scent. Much of this type of communication takes place unconsciously. We're all familiar with the sensation of just clicking with someone we meet. We call it "having chemistry." We're familiar with the opposite, too, when being with someone makes us uncomfortable. The explanation for both lies in our so-called nonverbal communications, including those transmitted by smell. Certainly crocodiles and snakes and lizards can signal with sound and smell and can therefore communicate. But crocodiles don't play. They don't have emotional lives in any sense that we would recognize. The content of the messages we send and receive is essentially different from that of reptiles. Our messages reflect the fact that we care.

Mammals imitate one another. Mimicry is an automatic behavior that helps to synchronize the emotional states of two animals to enhance communication. Paul Ekman, a psychologist specializing in the facial expressions and body movements involved in nonverbal communication, showed experimentally that if you spend enough time frowning, you will begin to feel sad. If you stand up straight and walk with a purpose, you will feel more confident. Working out makes you feel good for many reasons, but one of them is that it forces you into attitudes of confidence and power. A rock star jumps around and shoots his fists into the air for the same reason that Hodari, the gorilla in the Milwaukee Zoo, beats his chest. Expressing an emotion with the face and body causes you and those around you to feel that emotion. Mimicry makes it easier to feel another's emotions, which makes it easier to cooperate (or compete). With few natural defensive systems, creatures like us were more likely to survive when we cooperated.

The complexity of this emotional interchange, which we call empathy, represented a marked evolutionary change. For mammals, suddenly, there was a terrible new fate: that something should happen to your parents, your siblings, your loved ones, and most especially to your children. Caring meant that, for the first time in the history of life, there was a fate worse than death.

Two people who are communicating through these nonverbal channels will not only catch each other's emotional states; they will unconsciously synchronize their body movements, speech patterns, and even heart rates and breathing. It has taken high-speed cameras and other recent inventions to detect this system at work. Ekman pioneered this research, showing that complete shifts in facial expression—from a happy smile to a horrified grimace and back to a smile—can take place within a fifth of a second, too short to register on our conscious minds but long enough to influence our behavior. Two people who are communicating in this way will synchronize within as little as one-twentieth of a second. Elaine Hatfield summed it up in her book *Emotional Contagion*: "The number of characteristics people can mimic simultaneously is staggering." If you want to see this dramatically in action, simply watch children play together.

People who are engaged in synchronized conversation will mirror each other's body postures, crossing their arms or legs together or reaching for a drink at the same time. Once this synchrony is set in motion, it draws people together. They like each other more. The better the synchrony, the better the rapport. If we all perform the same physical acts, then, we synchronize. This is one reason why we like song and dance and exercise classes. It's why soldiers march. Drumming is another example of how people synchronize their emotional states. When the Spanish brought African people to Mexico in the 1500s to enslave them (the reason why many Mexicans are so dark-skinned*), they took their drums away to keep them disoriented and shatter their groupness.

Although everyone is different, most people not only synchronize but find emotional synchrony essential for what we call rapport. When two emotional systems are out of synch, the discomfort can be excruciating. Likewise, it's painful to synchronize with someone who is feeling awful. If you are receptive, you can catch someone's dread, depression, fear, or sadness. This effect can be devastating to a marriage.

People unconsciously synchronize in crowds. One of the most remarkable examples of this is the way people move on crowded streets without anyone directing their behavior. In his book *City*, William H. Whyte described how pedestrians maneuver on the sidewalks of New York. "With the subtlest of motions they signal their intentions to one another," he wrote. "There is a beauty that is beguiling to watch." Synchronizing with a crowd at a concert or sporting event can be a joyous occasion. And anyone who has seen a crowd stampede knows just how frightening it can be when synchrony goes out of control.

What I'm describing is a very high-level example of what appears to be a physical characteristic of many self-organizing systems. Steven Strogatz, a mathematician at Cornell University, is well known for his work on synchronization in all sorts of systems, from swarms of Asian fireflies that flash together, to the way cells synchronize their firing to produce the rhythmic, coordinated beating of the heart. At a talk he gave at the Credit Suisse First Boston Thought Leader Forum

*For more on this, see Mexican Fine Arts Center Museum (now the National Museum of Mexican Art). *The African Presence in México.* Chicago: Sheffield Press, 2006.

in 2003, Strogatz said that "synchrony breaks out suddenly. It does not build up gradually. It breaks out in a . . . phase-transition fashion. It has to do with positive feedback."

My first experience with positive feedback was with the vortex in a bottle, learning how molecules moving in a downward spiral can recruit other molecules to do the same until a transport channel is formed. Positive feedback can be seen in herd behavior, too. If a herd of cows is standing still, milling around, there is no direction to the movement. But if a few cows happen to move (or even look) in the same direction, others will mimic them and follow. The now-larger number of cows moving in the same direction exerts an even greater influence on the others, and there is a cascade of mimicry and synchrony, causing a change in behavior. The herd moves as one. If an energetic kick is added to the system, such as a loud noise or the scent of a predator, the herd will undergo a more abrupt type of change in its flow and will stampede.

Knowing where someone is looking is a very important tool of communication. It helps us read another individual's intentions. It helps coordinate how we relate in groups. Roel Vertegaal and Yaping Ding, two researchers at Queen's University in Canada, showed that if the people in your group look at you, you become more self-confident and talk more. Your brain counts up the number of times people look at you, and, as Judith Harris put it, "The counting device is one of the components of the status system; it counts eye-gazes the way a voting machine counts votes." When the number reaches a threshold, a synchrony signal initiates a new behavior: you speak. Animals make decisions by voting, too. Two researchers at the University of Cambridge discovered that African buffalo count eye-gazes as a way of voting on which way to move.

If I'm the parent of a teenager who suddenly looks weird because of a new hairstyle or clothing, I may wonder what's going on. It's part of the status system, working to accumulate more eye-gazes. You may argue that a well-adjusted teen wouldn't need to go to such lengths, and that may be true. On the other hand, at least I can understand that my kid is just working on getting status by making people look more often.

At times, we may find it useful to refuse to synchronize with someone. Police do this when they arrest someone or give out a traffic ticket. They don't want the other person's insides getting into them. That just causes empathy. They want entreaties to bounce off of them, so they tune down their affect as much as possible to prevent synchronization.

I once had to give a deposition in a lawsuit, and my lawyer told me to answer precisely what was asked and volunteer nothing more. It began something like this:

Q: Can you tell me your name?

A: Yes.

Q: What *is* your name?

And so on.

All of my answers were delivered with a completely blank expression and no vocal inflection. After about four hours of this, the plaintiff's attorney was screaming at me and pounding on the desk. During a break, I asked my attorney how I was doing, and she smiled and said, "Just fine."

There's a good reason why the other attorney was upset. Withholding affect, an essential component of the nonverbal communication system, is a terrible thing. It means that you're denying someone else access to your emotional system, which is the same thing as saying you don't care. It also means that the person is out of the group. In our origins as apes, inside-to-inside communications were essential for survival, and being out of the group often spelled death. Unless you could find another group, ostracism was fatal. Infants depend completely on an ability to read the mother's emotional system and count on her to read their own needs. Infants who don't receive attention and affection grow sick and may even die.

Withholding that essential kind of communication sends an ancient signal in the oldest language we have, a language that we can't hear but that we can feel in our bodies. It thus carries a message that we believe deeply, even without being conscious of that belief. Creatures that care need to see evidence that others of their kind care, too. Not receiving the cues that signal care, being cast out of the group, therefore feels like a death sentence. Who wouldn't be disturbed by receiving such a message? This explains why things can go so terribly wrong when one partner in a relationship employs "the silent treatment."

These tendencies—to safeguard the inside self, to synchronize through positive feedback, and to care—have exerted their influence throughout our evolution. They have given rise to our strong disposition toward groupness. All human groups develop customs, rituals, songs, dances, and rules that serve to synchronize the members and facilitate communication and useful relationships among them. By doing that, the group itself forms another important inside–outside relationship. If our membership in our group is disrupted, the effects can be very dramatic. Those who understand this can manipulate disenfranchised people to their advantage.

When groupness is disrupted, most people immediately seek to join or form another group. That makes them far more susceptible to cults and mass movements. Eric Hoffer connects a loss of self-esteem with the hopes for the future that such groups hold out to people. "Those who see their lives as spoiled and wasted crave equality and fraternity more than they do freedom," he writes. "A man is likely to mind his own business when it is worth minding. When it is not, he takes his mind off his own meaningless affairs by minding other people's business." Hoffer holds up Jesus as a good example of someone who knew how to destroy a sense of groupness, specifically for the purpose of making people susceptible to being recruited into a mass movement. Jesus often preached against the family, which was the basic group at the time. ("Jesus minced no words," Hoffer wrote, and then quoted the New Testament. This is from Matthew 10:35–36: "For I am come to set a man at variance against his father, and the daughter against her mother, and the daughter in law against her mother in law. And a man's foes shall be they of his own household.") Jesus found most of his converts in the city, not the countryside, because the cities of his time were already home to many disenfranchised individuals who had lost their traditional groups. The new followers of Jesus derided those who lived outside the larger cities, because those groups were stable, their present state assured, and they were less easily influenced by hopes for the future. The ones in the villages were called *pagani*, from the Latin word for village. The ones out on the heath, similarly stable in their groups, were called heathens.

Hoffer also points out how debilitating the effects of losing your group can be, and how groupness can work to protect a people, too. The European Jews had had their groups shattered before the Nazis began carting them off. They did not put up concerted resistance to being taken to the death camps. Once the Zionist immigrants from Europe were in Palestine, the British expected them to be pushovers for colonial rule. But it was the English who were sent home with their tails between their legs by a newly ferocious and closely knit group that had rediscovered the power of keeping the inside separate from the outside. The group had found new ways to care.

The phenomenon of the mass movement grows out of a natural and ancient feature of human behavior. It is intimately bound up in the ways we communicate. The group sends a powerful message. It can completely change the way a person behaves, sweeping away what we think of as a stable moral and ethical landscape in an astonishingly short time. (As the Stanford prison experiment, the Robbers Cave study, and Abu Ghraib prison have shown, what we consider to be aberrant behavior is fairly easy to redefine.) That is because the message is sent not in words, not in propaganda, but in the secret language of nonverbal communication. Charismatic cult leaders know two very important things. They know which people to recruit. They recognize those who are good recipients of the message. (Hitler, for example, considered Communists prime candidates for his National Socialist movement, even though they were seen as the enemy. They were already zealots, which qualified them for any mass movement.) And they also know how to get inside another person's emotional system and rewrite what's in there to change the object of caring.

Groupness and the natural functioning of the emotional system are neither good nor bad in themselves. They are simply features of our makeup, and can be used for our benefit or to do harm. Knowing what goes into determining what we care about and what we're willing to act upon, however, can help shape our future. As we've seen, in biological and even in nonbiological terms, all these systems use memory to predict, plan for, and influence future outcomes. But, as we've also seen, these systems work on the principle that the future is going to look like the past, at least in its general outlines. Animals would not have evolved to predict the future this way unless future events really did bear some resemblance to past events. In other words, if we can know what the general shape of events in the future is apt to be, we'll be more successful at choosing a course of action.

Fourteen

The Climax Shape

I began my research with a simple question: Why do smart people do stupid things? I approached it from the psychological level, then the neurological level. I dug down and found hints that suggested one answer, but then I'd detect another layer beneath that. I'd explore that layer and gradually come to grips with it (or not), only to spy another layer beneath that. The more deeply I explored, it seemed, the farther back I had to go. If neurological processes help determine how we behave, then do the chemicals that make up the nerves also exert an influence? What about atoms? Does where we came from shape all we do? If so, then how far back can I go? How basic can it all get? First principles always seem like a worthy goal to me. Let's go as far back as we can go and see where it gets us. Let's go all the way back to the beginning of everything.

Edwin Hubble, after whom the Hubble telescope was named, discovered that galaxies that are twice as far from us are moving twice as fast. He also observed that everything is expanding in every direction from every possible point of view. The only conclusion to draw was that everything we observe in the universe started at a common focal point and expanded outward suddenly. That event, which we call the big bang, began between 10 and 20 billion years ago. The most common estimate hovers around 13.7 billion.

The second law of thermodynamics meant that the energy which was there at the beginning had to spread out. The initial temperature of the big bang can be described by the number 10 with thirty-two zeros after it in the Kelvin scale. That number is greater than the number of seconds since the big bang, which is a mere nineteen-digit number. At first the universe was nothing but matter and antimatter rushing apart. But for whatever reason, there was a bit more matter than antimatter, about one part in a billion, so some was left over. That leftover matter cooled, turning into such particles as photons, neutrinos, electrons, and quarks,* the building blocks of the matter we know. At some point before one-hundredth of a second had passed (perhaps much earlier), as the temperature cooled, the weak, strong, and electromagnetic forces, which had been one, split from one another.

When about three minutes had elapsed and things had cooled down a bit more, protons and neutrons, the components of atomic nuclei, formed. They found each other and stuck together, forming hydrogen and helium and traces of lithium and deuterium, and that's pretty much the way things stood for the next 300,000 years.

All of this was taking place as a result of a struggle on a stage that was set by gravity, which pulls things together, and quantum mechanics, which keeps things apart. On that stage, the second law of thermodynamics acts to move energy, and in its struggle to do so we get something unexpected: order and information. The structure we see represents the interactions between the driving

*Murray Gell-Mann predicted and named the quark, which is thought to be the ultimate subatomic component. Quarks were independently predicted by George Zweig when he was a graduate student at Cal-Tech.

force (energy wants to move) and the nature of space and matter through which it has to do its job. As with the vortex in the bottle, the driving force has to push stuff out of the way, even while gravity is trying to pull it together. This movement of energy in the direction of increasing entropy is the same general tendency that started metabolism going on the ocean floor in the early days of the earth.*

The play taking place on that stage has a distinct shape, one that is found throughout the universe:

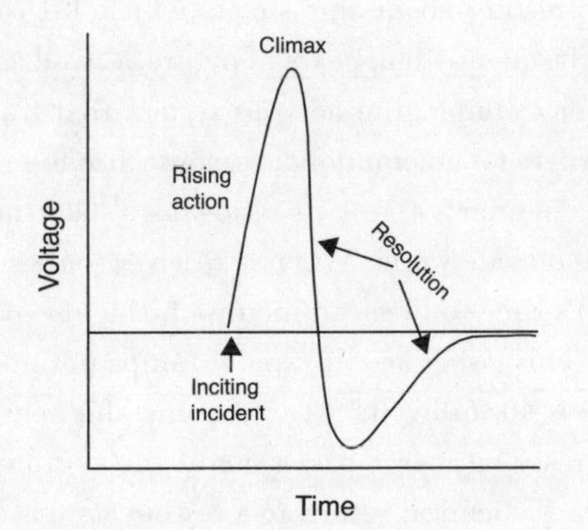

There is an inciting incident, rising action, a climax, and a resolution. We don't know what the inciting incident in the big bang was, but we assume that there must have been one. The rising action is the spreading out of all that energy, climaxing in the creation of the matter and the four forces that would shape our world for the foreseeable future. We are now in the resolution phase, somewhere on the backside of that curve.

The curve also represents the shape of an emotional response. If you're walking down a quiet path in the jungle and a lion rushes at

*I'm indebted to Eric Smith for these explanations. As he told me, "The difficulty of handling such combinations of stage and energy conditions is what has left questions of this type beyond the forefront of statistical physics."

you, teeth bared, you will have a response that has the same shape as that graph. The inciting incident is the sight (sound, smell) of the lion. The rising action involves all the chemical and physiological changes that you undergo that allow you to run or fight. The climax comes when you reach safety or when the battle is decided in victory or defeat. The resolution unfolds as you catch your breath, as your heart rate slows (or your corpse cools off), and as the chemicals that made you respond the way you did are gradually metabolized until you feel normal again.

What's interesting about this shape, which I'll call the climax shape, is how frequently it appears in nature, as well as where it does not appear. For example, the nervous system that transmits all the signals and orders for an emotional response also follows this climax shape. When a neuron is at rest, its inside has a stable negative charge relative to the outside world. When it receives some signal from the outside that it's time to fire—an inciting incident—positive sodium ions rush in. This rising action rapidly ramps up until the positive charge inside is 40 millivolts. At that point, the neuron reaches its climax and fires. The charge passes out of the neuron and down the axon pathway. A chemical return to a resting state is the resolution. The diagram above happens to be a schematic illustration of a typical neuron firing.

The climax shape isn't just limited to biological systems, either. It can be seen in phenomena from weather systems to earthquakes to the structure of movies. Drama has the same shape as an emotional response, because it must elicit one to be effective.

To find our earliest beginnings, we must go far beyond the age of the earth. We're made of dead stars. The sun is a typical star, if such a thing exists. A star is a lot of hydrogen being pushed together by gravity with such force that the nuclei of the hydrogen atoms fuse into helium in a thermonuclear reaction, the same reaction that takes place when a hydrogen bomb explodes. The temperature of the sun is roughly 27 million degrees (about 15 million C) inside and about 10,000 degrees on the surface (about 5,538 C).

The helium produced by stars, which began forming about a billion years after the big bang, is useless in creating life. That's why we had to wait for stars to start dying before we could be born. When a star begins to use up its hydrogen, it starts to collapse under the push of gravity. The temperature and pressure rise to the point where three helium nuclei are pushed together to become one carbon atom. The crushing force of gravity creates a kind of adding machine.

Here's how it works. Hydrogen, the simplest atom, is composed of one proton (positive particle) and one electron (negative particle). When it's crushed hard enough, two of those protons are bound together with two neutrons (neutral particles). Opposite charges attract, so that pulls in two electrons, and the new structure contains two of each particle—2–2–2, which is helium. Gravity never rests. It pushes three helium molecules together hard enough to produce 6–6–6: carbon, which has six each of protons, neutrons, and electrons. Pushing carbon together adds two more of each, producing 8–8–8, which is oxygen, the third most abundant element in the universe after hydrogen and helium.

Hydrogen, carbon, and oxygen are the most promiscuous elements in terms of their ability to form compounds with other atoms. It seems no coincidence that these are the chemicals of life, making up 93 percent of you and me. Oxygen can combine with hydrogen to make water. Half the material in the earth's crust is oxygen. So there, in these few simple steps, we have almost all we need to sustain life. The earliest metabolism, the citric acid cycle, uses these three elements—hydrogen, carbon, and oxygen—in a cycle of positive feedback that results in reproduction of large molecules. From that cycle, amino acids, peptides, and proteins can be made, because carbon has the unique property of being able to make long chains, which are necessary for life. As in other systems that we've seen, what starts out as a simple quantitative change is somehow transformed into a qualitative change. Addition leads not merely to more of the same but to new worlds.

The adding machine of the star continues to crunch numbers (and nuclei) for as long as there's enough material and pressure. If there is enough mass to start with, the process will produce all the elements right up through iron, with twenty-six proton-electron pairs and thirty neutrons.

If you look up into the night sky at the right shoulder of Orion, you can see where this process is happening. Betelgeuse (Alpha Orionis) is the red star you see, and it's dying. It's a big one, too, likely capable of producing all the elements necessary for life. If Betelgeuse were located where the sun is now, its outer edge would be about where Jupiter is. That makes it one of the larger stars, some 600 times the size of our sun and 60,000 times brighter. The reason it seems faint to us is that it's more than 400 light-years away. Look up there tonight if you can. When the light you see was produced, Caravaggio was busy painting *Salome with the Head of St. John the Baptist.*

Right now, carbon and oxygen arc probably being made inside Betelgeuse. As the star ages and contracts, it will push together nuclei to form neon, magnesium, sodium, silicon, and eventually iron, another material we need in order to live. Because iron is so stable, Betelgeuse will eventually run out of matter to compress. It's these compressions, transforming one material into another, that allow the star to spew out all the energy that we receive as light. If the star is no longer compressing material, it's not putting out enough energy to hold back the crushing tide of gravity. At that point, the star collapses and explodes, becoming a supernova and casting off all this new material (and more energy) to be carried away to form comets and asteroids, which might contribute to sustaining a future race that could organize itself out of those materials.

After the supernova ejects all that useful matter, Betelgeuse will contract into a mass about one and a half times that of our sun, crammed into a 12-mile-wide ball spinning at about 600 revolutions a second. The pressure inside will be so great that the protons and electrons will fuse and become neutrons. The entire cycle is circular, in a sense. The material in Betelgeuse after it turns into a dead star will be much like the state it was in about a second after the beginning of the big bang. As my father used to say, "Nature loves a circle."

Like the big bang itself, the life cycle of Betelgeuse exhibits the climax shape, the same as a neuron firing or an emotional reaction: inciting incident when gravity pulls enough material together to ignite the fires of fusion, rising action in the increasing complexity of the atoms it creates, climax in a supernova, and resolution in the scattering and cooling of all that matter as it condenses once more into new forms, as comets and asteroids and dust and debris are pushed together to form planets like ours, some of which will contain all the matter and energy necessary for life.

Almost half a century ago, I stood in a medical school lab, watching my father swirl a bottle to produce a vortex that allowed water to drain more efficiently down the hole, and I began wondering about that structure and its central meaning. Its meaning was that it was transporting energy to increase entropy in the best way it knew how. It was running the way the universe runs. At the center of the galaxy where we live is a black hole. Nearby stars caught in its gravitational field orbit that vortex at three million miles an hour, ten times the average speed of stars. Black holes are increasing entropy, too, but on a cosmological scale that is hard to fathom. A black hole with the mass of only three of our suns would have entropy equivalent to the number one with seventy-eight zeros trailing off.* That's for only

*As mentioned before, entropy equals the heat added to a system divided by its temperature. Entropy is measured in "joules per kelvin." Any number with seventy-eight zeros before the decimal point is very large, even on a cosmological scale. For comparison, the estimated number of electrons in the universe is a number with seventy-nine zeros.

three suns. Here at home in the Milky Way, our local black hole is equivalent to 2.5 million suns. Out there in the universe, there are black holes that are billions of times the mass of our sun, and they, too, are draining everything around them just as fast as they can. The universe is home to vortex structures at all scales.

If you've stayed with me until now, you may be wondering how we've gotten so far out on this limb. Here we are, somehow, wandering among galaxies and black holes. What does this have to do with grabbing rattlesnakes and the variety of breakfast cereals in the supermarket? Way back in Chapter 1, I mentioned that the anthropologist Colin Turnbull once took a tribesman named Kenge out of the Ituri Forest and showed him a herd of buffalo on the open plain for the first time. Kenge refused to believe that the buffalo were not simply small insects, because that was the only mental model that matched what he was seeing. I wanted to get out of the forest, so to speak, to snatch us away from our everyday cues, which activate the same old models and scripts. Way out here in this rarified air, we're forced to take a new look at things. If we can see that a dying star was required to crush together all the right materials to make a box of Quaker Simple Harvest Maple Brown Sugar with Pecans Instant Multigrain Hot Cereal, then our world will never look quite the same again. If we can see that we are united with the laws and ways of the universe—from the way our nerves fire to the way we process energy—then we may develop a more accurate view of nature and of the influence it has in our lives. This new view may show us that we are not set apart. We are in the thick of it. And it is only by changing our view of things completely that we will have any hope of changing our behavior.

Fifteen

The Guest Star

Lucy and the Laetoli woman knew they were in the thick of it. They most likely didn't have the ability to reflect very analytically on what that meant. Theirs was not yet a world of symbols through which they could manipulate and rearrange the world to a very great extent. Long after they and their kind were dead, by about 50,000 years ago, people who were more like us had begun to decode their world, bending it to their will, and were even trying to understand the universe. I wanted to find a way to connect with their efforts to place themselves among the stars and galaxies.

I had heard of a spot out west at about the 120th meridian near the 42nd parallel where Interstate 84 veers up toward Portland and Interstate 80 dips down toward San Francisco, leaving a great swath of land nearly untouched by recent depredations. It contains the most empty space in the contiguous U.S.: hard, unpeopled desert. A place of illumination.

My friend Jonas Dovydenas and I flew a small high-performance plane out there to a place near the Alvord and Black Rock deserts. We took turns flying from Chicago for eleven hours to reach Elko,

Nevada, where we spent the night. The next morning we rented a four-wheel-drive vehicle so that we could go deeper into the land. We drove as far into the desert as we could and then got out and walked.

We crossed a shattered lava dome, immense fields of jagged iron-red stones. We passed through alkali and pumice flats, and descended ravines of broken volcanic glass. At each high vantage we found eagle droppings, burned white by the sun. And yet nothing visible grew there but a few burnt-over sagebrush bushes. So where did the eagles find their prey? As we topped the next rise, the smell hit us. Here was the answer, the secret of that land: wherever you pour water onto stone, life explodes in a profusion and diversity that is both mystifying and inevitable.

We hiked down out of that seemingly dead world of dust and rock to a slow-moving brook, rich with sedges and reeds and grasses, watercress and fern, fairy rings of flowers in pink and yellow and purple. The closer we came, the stronger the perfume of life, until we saw iridescent bottle-green dragonflies buzzing among reeds of biblical antiquity. Sign of fox, cat, rabbit. Fish swam in the cool waters in this place where a man could die from the heat in a day's time.

We pressed on until we came to a dirt track, which we followed through low hills. Around a bend, we found a cabin, a few tents set up nearby, a dog tied to one of them. Someone was there, we thought we knew who. We spotted a blue canopy half a mile away against the side of a cliff and began crossing the rugged terrain. An hour later, we found what we expected: archeologists. Hence the tents back at the cabin. Like all the dreamers who are seduced into that land, they had a rugged beauty about them. Like prospectors, they had an air of almost giddy contentment as they worked in the hot sun, sifting the dirt that seemed hammered into every pore of their grinning faces. Kelly McGuire, one of the owners of the Far Western Anthropological Company, had brought along his nine-year-old daughter, Chloe, who was asleep with her head on a folding table in the shade of the blue canopy. Another archeologist with the company, Kim Carpenter, was pregnant and seemed about to burst out of her dust-covered denim overalls. She showed us a perfect arrowhead, which someone had chipped out of green translucent stone thousands of years ago. In a way, these archeologists were doing just what our ancestors had always done: sifting, pounding, stirring—organizing repetitive motions of the hands to shape the thoughts in the brain.

Archeologists are drawn to the area because it was so densely populated until perhaps one thousand years ago. On Kim's advice, we met up with another archeologist a few days later and headed out on a boulder-strewn track toward the Surprise Valley near Massacre Lake, passing through fields of wild crested wheat, where feral horses ranged free. We had two other vehicles from the Bureau of Land Management following our truck, because if the road killed two vehicles, there'd still be one left. We didn't want to try walking out. The nearly impassable road terminated in a stand of ancient twisted junipers. This was where Kim had told us to go to find what she referred to as "a two-mile-long art gallery" of ancient petroglyphs.

We hiked a dry wash to a crumbling rock wall. Sheets of stone, which had sheared off and shattered on the declivity at its base, made for a rugged walk. Chipped into the dark brown rock, one after another, were pictures of the world that an ancient people once saw. Some of them showed the looping course of rivers in the valley below, now vanished. One depicted a man-sized fish hanging head down, as if freshly caught. The archeologists from the BLM explained that this land had once had plentiful game and water and vast fields of grasses to provide a harvest of wild grain. No more. The hair on the back of my neck stood up, as I realized that I was seeing the world through the eyes of someone dead these thousand years.

One of their grinding stones was wedged in a crack in the wall. It was not so different from the stone metate I'd seen my great-grandmother use to grind corn for tortillas when I was four or five years old. We also found one of their pointed hammer stones, which they'd used to chisel the petroglyphs. Those tools had been cached there for at least a thousand years, awaiting the nomadic tribe's return. But one spring, the people didn't come back.

We came to an image that Kim had mentioned, the one that I most wanted to see. It was striking because it was so unlike the others. It showed a quarter moon in opposition to a bright star. There were hash marks beneath, as if someone had been counting. Chinese literature describes a supernova that was visible in full daylight in 1054 AD. That climax, which created the Crab Nebula, would have been visible from the spot where we were standing. The hash marks ticked along at regular intervals beneath the image and then stopped. I counted them. There were twenty-three.

"How many days was the supernova visible?" I asked.

"Twenty-three," the archeologist told me.

On July 4, 1054, the Chinese observed what they called a "guest star" as it appeared in the constellation Taurus. The Anasazi and Mimbres in Arizona and New Mexico also depicted the event both in petroglyphs and on pottery. In some of the images, the counting is done by showing a star with twenty-three rays. On the morning of July 5, the supernova would have seemed to be riding very close to a crescent moon as viewed from where we were standing, just as the picture on the wall showed us.

The petroglyph we were looking at was simply the sign of hand and eye and brain, the mark of a human being trying to know his world and keeping track of an unusual occurrence in it. Language is an extension of our emotional system, which stores and labels everything for future reference in an effort to help keep us alive. It is a powerful time-shifting device, a code used to extend our memory system far into the future. And this picture before me was nothing less than the written history of a strange event and a person's attempt to grasp it. Emotional systems, as we've seen, engage each other and synchronize using the classic climax shape that the universe seems to like so much. Oral tradition, art, imagery such as what we found on that rock wall, and later written language, have given our emotional systems an almost limitless reach by generating what we call history. The person who sent the message about the supernova had reached across millennial time and plunged his hand into my heart to re-create that special shape in me.

What we were witnessing in this and all the other art across the wall was the direct result of small but very important evolutions of the hands of our ape-like ancestors, which were gradually transformed into what we begin to see in fossils from a few million years ago: a genuine human hand, evolved so that the fingers and palm were supple enough to hold oddly shaped objects, powerful enough to hammer, and so that the hand and wrist were capable of absorbing the shock.

We found a place to climb the rock wall and came upon more recent signs of man. A bronze disk had been cemented into a rock near the precipice. It was inscribed "Wallace L. Griswold" and beneath, "1918–1982." Penny Carmosino, one of the archeologists, knelt down and picked up a fragment that I thought was a porous white rock. "This is part of a human skull," she said. I noticed that Griswold had been born in the year that the Spanish influenza infected a third of the world's population, the most recent extinction event in our own species.

Penny was turning over pieces of what I thought was black flint. "This is cinder, see?" Tarry black matter that had once been Wallace. Penny looked out over the valley and said, "Someone cremated him and put his ashes here. Wallace must have loved this place." Penny understood how Wallace's soul had been captured by that place.

As I held a piece of Wallace Griswold's skull in my hand, I thought, if this is all that's left when we go, then we had better visit places like this, if only to get our bones used to the emptiness where they will come to reside anyway—and for far longer than we had possession of them. I also thought how remarkable it is that we have found any fossils of our ancestors at all. I would have walked right past these fragments, never knowing what they were. It's no small thing to do real archeology. Someone must come with the intention of finding, with a theory of where to look, with a special skill at seeing, and with a deep feeling for where to point the human eye.

We hiked back over ground strewn with evidence of arrow points, chipped-off bits of obsidian called bifaces. Obsidian doesn't occur in that area, so someone had to bring it there. As the sun reached a certain low angle, the whole land lit up from reflections off the broken obsidian. We could see it glittering all the way to the horizon. It was eerie, because it so clearly demonstrated how teeming and busy the place had once been, in stark contrast to how deserted it was now. And that hypnotic light, which came but once a day, and mostly when not a single living soul was here to bear witness to it, now formed the last luminescent bones of the people who had called this place their home in whatever language was theirs, with whatever word or sign.

Jonas and I decided to stay out in the desert on our last night in the area. With no significant artificial light for perhaps 100 miles around us, it seemed like a good place to view the workings of the universe in the night sky. We flew north-northwest out of Elko up into the flank of the Steens Mountains and spotted white patches on the earth at a distance of 50 miles. They were so brilliant that at first we thought they must be water. As we drew closer, we saw that they were dry lakes.

We surveyed several small lakes and selected one that looked promising. A quarter mile would have done nicely, and half a mile would have been extravagant for landing the small plane. This nameless place was about nine miles long and clear of obstacles except for some greasewood bushes cutting across it at an angle. But we didn't know what the surface of the lake was made of. If it was sand or loose dust, we'd bury our wheels, and the plane would tumble end over end. That would be bad. Again and again, we flew over it, looking for some sign.

As we made our third pass, angling up into the air and circling around like a crop duster, I wondered what we might see in this featureless place that would tell us whether the lake was deadly powder or a hard surface we could land on. We came in just a few feet off the ground. Jonas flew, slowing the plane, as I looked out the side of the canopy for a sign that we'd be willing to bet our lives on.

Craning my neck to look ahead, I saw at last the orderly symbols that we both knew would characterize a hard, dry lake bed. Neither of us had discussed what we both knew: that since at least the big bang, nature had been trying her best to make spheres, as she had done with earth and moon and stars. Those spheres are made by gravity pulling inward from all directions, but spheres can be made in other ways. The surface tension of water makes droplets spherical (though they are sometimes pulled out of shape by gravity or the friction of the air as they fall as rain). But spheres are good. They are minimal shapes. And nature will always settle for quick and dirty approximations.

"Cracks," I said. "Hexagonal cracks."

"What?"

"Nothing. Land the plane."

He put the gear down and dropped more flaps, and as the airplane settled in, we felt the wheels touch a surface as smooth and hard as concrete. We both let out our breath as we rolled out toward the line of man-high bushes. I turned around and saw a tower of white dust gaining on us.

It was a hot and windy afternoon as we tossed our gear out and stepped down from the plane. As we set to work making our dry lake camp, I hammered a stake into the ground to hold our tarp against the wind. It was like trying to drive a nail into concrete, and the hollow, stubborn "tink!" it made as I struck the metal put me in mind of the person who'd made the image of the supernova on the wall of petroglyphs when this lake was teeming with fish. He had been watching a star spew out the materials needed to make someone like him. And all of these things, as I now saw, came from the same processes acted out on the same stage, and developed with the same climax shape.

The hexagonal cracks that Jonas and I saw were the result of water evaporating. As the soil contracted, it shrank. As the water went away, the soil became more solid, and the internal stresses grew until it couldn't shrink any more without cracking. Cracking in a hexagonal pattern reduced the most stress with the least amount of cracking.

Hexagons, like spheres, are found abundantly in nature. Snowflakes are hexagonal, and many atoms are organized in hexagonal crystal structures, including hydrogen, helium, carbon, and nitrogen. Bees use the least amount of wax by making hexagonal cells in their honeycombs.*

No one can say exactly how life first formed on earth, nor even if it did, since some people believe it fell here from somewhere else. Some scientists believe it may have come from Mars, for example, while others argue that Mars never had enough of an entropy gap to do the job. Another theory involves life forming in clay around the type of hydrothermal vents discovered by Robert Ballard. The clay solves a problem: even in an ocean that contains a source of energy and all the right molecules for life, there's no mechanism for organizing them and getting them into place so that they can mate with each other to form the more complex chains that spell out the vocabulary of life. Random mating makes the right combinations statistically unlikely. But clay, which is made up of tiny crystals, can trap molecules of a certain shape and hold them in a pattern. Water can pass through the clay, carrying other molecules with it. In addition, clay provides a stage on which a lot of materials can come into close contact with

*For those interested in learning more about processes that produce hexagonal transport channels, a Web search of Rayleigh-Bénard convection or Bénard cells will turn up abundant resources.

one another. Clay has a very large internal surface area. So clay can act as a catalyst for certain reactions, positioning the molecules in just the right way to cause the correct chemical reactions to occur. The trouble with this theory is that the right kind of clay may not have existed on earth when the earliest life formed.

A more compelling theory was put forth by Günter Wächtershäuser, a German chemist who built on the work of Helmut Beinert of the University of Wisconsin at Madison. Beinert first observed in 1997 that all life-forms contain proteins with iron and sulfur in them. Wächtershäuser demonstrated that when there's iron sulfide in the environment, it works to catalyze very simple molecules, such as carbon monoxide, to produce much more complex ones. He proposed that iron and nickel sulfides provided both templates and catalysts, as well as a source of energy, for the first biological molecules.

At the Geophysical Lab of the Carnegie Institution, George Cody, Robert Hazen, and others have been following up on Wächtershäuser's ideas, using high temperatures and pressures similar to those in the region of hydrothermal vents to synthesize pyruvic acid and citric acid in the presence of iron sulfide. Both acids are critical to metabolism in living things. Hazen has shown how pyruvic acid, when put under conditions similar to those at hydrothermal vents, would produce all the right organic molecules for a cell-like structure. He has produced such vesicles in his lab.

Knowing how life came about opens the possibility of beginning to sketch a line from the big bang to us. I don't think we are at a point where we can draw that line just yet, but we're at least beginning to learn what some of the steps will need to be on the staircase we'll have to climb to get there. That staircase will lead to Lucy and the Laetoli woman, and through them, it will lead to tribes and cities, pyramids and coliseums, to traffic jams and global warming. It may one day illuminate the mystery of how nature could make a man who makes a supermarket.

To do that, scientists studying the origins of life are attempting to form a bridge, step by step, between inorganic and organic chemistry—between rocks and life, so to speak. Some researchers, including Eric Smith and Harold Morowitz, now believe that the metabolism of the citric acid cycle, described earlier, came first and formed the basis from which more complex organic chemical processes could take shape. Only then could we have what we call life.

That citric acid cycle alone would define an inside–outside relationship, a climax shape, and provide a transport channel for energy (electrons from hydrogen atoms). Add to this chemistry the formation of a more well-defined inside—a cell, not a simple step—and you have life as we now know it. Where could such a cell come from? Through work by such scientists as Wächtershäuser, Cody, Hazen, Morowitz, and Smith, it's beginning to appear that the first living things on earth may have formed by a catalytic process in the bottom of the sea while being heated from below by the molten rock in the earth and supplied with high-energy electrons from hydrogen. After seeing the organized structures of a dried lake bed and Hazen's vesicles, it should come as no surprise that, as the surface of the earth first solidified, all sorts of organized structures were created widely and at all scales by a variety of processes that transported energy. Although it makes sense that metabolism came first, an accidental "self"—an enclosure such as Hazen made by starting with pyruvic acid—might have provided a more efficient environment to sequester the right chemicals for assembling complex molecules. Metals with their promiscuous electrons might have been trapped in these proto-cells and may have served to catalyze other chemicals and fix carbon into forms suitable for life. And that could explain why we find microbes no matter how far down in the sea floor we dig. The great tube worms that Ballard saw don't have an intestine for digesting food. They have an organ that is full of those microbes, which oxidize sulfide to supply a missing part of the worm's metabolism. We have intestines full of microbes, too. And as we've seen, microbes that lived independently long ago now thrive as mitochondria.

Matter and energy have a general tendency to follow certain patterns, such as making things round and creating dynamic systems with an inside and an outside. That physical tendency, operating in an environment where the type of biological chemicals created at the Carnegie Institution are available, may suggest a way that metabolic processes eventually became sequestered inside a cell. Once inside, that arrangement would be favored by natural selection, because the boundary concentrates the right chemicals for metabolism and keeps the wrong ones out.

G. Evelyn Hutchinson, a zoologist at Yale University famous for his work in helping to invent the study of ecology, viewed communities of organisms—ecosystems—as individual organisms with unique metabolisms. James Lovelock, who pioneered the Gaia hypothesis of the earth, wrote that the subsea microbes discussed earlier "were the whole biosphere for three billion years before multicellular organisms like us and trees came on the scene." The Gaia hypothesis holds that all life on earth behaves like one big organism, maintaining itself by controlling the environment of the earth.

It may be useful to view the earth as a single integrated self-organizing system attempting to optimize itself to increase entropy. Being subject to chance, geometry, and the constraints of local conditions, it does not reach anything that could be called an ideal state. Life is one important part of this system, and it does seem to exert a regulating influence in some ways. For example, a scientist from the U.S. Geological Survey discovered that microbes appear to control the flow of water into and out of aquifers deep in the earth, much like valves operating in a circulatory system. And of course, as the earth evolved, eventually algae produced enough oxygen to regulate the atmosphere so that our more familiar forms of life could come into being, around 1.9 billion years ago. From there life began to grow into more and more complex structures, eventually evolving to what we see today. I suspect that it did so not because each new form was individually that much better at dissipating energy but because each of those changes made the entire system—a local ecology or the whole earth itself—more effective at producing entropy. An earth with all ecological niches filled with life produces more entropy faster than an earth with empty niches or a dead rock planet like Mars. But just as each individual lifespan follows the climax shape, so, too, does all life on earth, going from inception to proliferation to a peak in population (and energy transport) followed by extinction.

Species are packed together into their ecological niches in a pattern reminiscent of fractals. Smaller species see resources at one scale, and larger species see them at another scale. A mosquito in the jungle will make a meal out of a drop of my blood. A tiger will want to eat all the rest. If life is viewed from the point of view of physics and energy, this makes sense. Recall that the problem of filling space with a distribution network was neatly solved in the circulatory system by creating a quasi-fractal shape. The ability of animals (including us) to get resources is also a spatial problem. Collectively they must exploit the resources in their habitat and fill that space, in effect, as completely as possible. That tendency to move the most energy at the lowest cost, along with the constraints of geometry, will naturally drive them in the direction of a fractal solution to the problem.

That this type of structure occurs in ecosystems tells us something interesting. It tells us how interconnected the structure of life is. While the Gaia hypothesis may or may not be correct, it can provide us with a helpful metaphor. Viewing the earth and life on it and the ways of the universe around it in a more integrated way may help us to accept how intimately interwoven life and these processes are. And that realization can help open our eyes to the fact that we really might do ourselves harm if we don't care about important things, such as the oceans, the air, and the other life-forms that may seem irrelevant to our activities but are actually crucial to our survival. A failure of that sort would represent a remarkable turn of events, in which the smartest creature on earth somehow managed to do the dumbest thing possible.

On the night that Jonas and I spent out on that dry lake bed, I made a fire from the thorny greasewood bushes. The sun withdrew behind the mountains as the wind picked up and made the flames roar like a smelter, and drove orange cinders in wild arcs across the lake. I watched them bounce along until they dimmed and vanished in the distance. The vast and empty land stretched away from the fire's nervous glow, and the tarp snapped in the wind, as we sat and watched Cassiopeia rise across the misty wheel of the galaxy. Mars and Jupiter towed a hollow moon up through a far deck of stratocumulus. Holes in the clouds sent spotlights down across the white land and made a death's head of the moon, until it breached the wall of cloud, luring mountains out of the dark. Above the cold moon, the Milky Way faded, and a million worlds winked out. The desert glowed all night with an eerie chemical light.

Sixteen

Land's End

At the end of my journeys, I found myself in Hawaii hiking the Na Pali coast. As I stood on a rock in a rushing river that cascaded down the cliffs to the sea, I watched as a beautiful young woman walked out of the jungle completely naked. She had long brown hair and smooth tanned skin. In one hand she held a shredded red sarong. In the other, she clutched two large avocados. She picked her way across the river rocks, wrapped herself in the sarong, and settled onto a boulder by the bank. She ran a long fingernail around one of the avocados, splitting the skin, then twisted it in half, revealing the bright green meat inside. Then she ate delicately, scooping the green fruit out with two fingers, licking it off with her pink tongue, and taking occasional dabs of it to smear onto her legs and arms and face for the oil it contained.

She was a kid from the mainland, and had vanished into the wilderness more than a year before. She had been living off the land in the Kalalau Valley ever since. She wasn't all that unusual. A lot of people go missing here. Some of them become confused on what was meant to be a brief hike and are probably dead by now. Some become enchanted and begin to think that it would be a good idea to experience what it's like to return to our origins. She was a modern-day Lucy.

I had studied satellite photographs of Hawaii before heading out here. It's in the middle of nowhere, 2,000 miles from the nearest land, and it looked like a good place to see the workings of the world. The Hawaiian Islands rise as a series of tiny blips of land starting at Midway Island, and surge up out of the sea to almost 14,000 feet. From space, they look like a code, a signal, a message. But there's more. I saw that the sea was crisscrossed with these intersecting signals. Something tremendous was going on here, and I had a hunch about what it was. I would have liked to go to the bottom of the sea, like Ballard did, to see the factories on the roof of the world. But this would have to do.

I took a boat trip around the island and saw the black and rust-colored lava cliffs of Na Pali tinted faintly green with vegetation. Valleys cut regular patterns where the water drained between the cliffs. As we approached Kalalau Beach, I was watching the cliffs, looking at the boulders along the shore, and thinking that we were perhaps 100 yards out. Then I saw dark points on the sand and realized that they were tents. My senses were jolted, as the optical illusion re-formed and my mental model was wiped out.

It turned out that we were half a mile offshore and the cliffs rose the better part of a mile straight up. And when the scale of the place snapped into view at last, it took my breath away.

Here's how the Hawaiian Islands formed. Deep beneath the ocean, molten rock punched a hole through a thin spot in that portion of the earth's crust known as the Pacific plate. The lava flowed out and built up a mound that rose out of the water. As the volcanic mound cooled, the tectonic plate moved on and carried it away. Then the heat punched another hole in the plate, formed another dome; and that island moved away, too, as gravity pressed down and slid the plate along against the thermodynamic pressure from beneath. I began to get the picture of a conveyor belt, continuously moving, its action driven by gravity from above and heat pressure from below, the forces competing and translating into a sideways, sliding motion. And as it moved, it left these volcanic islands behind. The weight of each island sealed the hole that had made it. Then, when the island moved off, the hot spot punched a new hole and began again. Over time, the archipelago stretched from the Kure Atoll just off Midway Island to the big island, Hawaii itself, reaching as far as the distance from Houston to San Francisco.

Some of the biggest landslides ever to occur on earth burst forth into the sea when pieces of Kauai sheared away, leaving the cliffs of Na Pali that I could see from the boat. Some of the debris tumbled 150 miles out into the sea, pushing a towering tsunami ahead of it. The heat of the planet, straining to get to the cool of outer space—to come to equilibrium—had created the structure of a cone. It was like the sand pile, but the energy and material came from below instead of above. Gravity brought the pile down. As with the sand pile, there were many small collapses, fewer intermediate ones, and fewer still of the giant slides that crafted these cliffs. The collapses occur in a fractal pattern with each describing a climax shape.

But the destruction wouldn't end there. The whole archipelago, riding on the conveyor belt of the Pacific plate, is sliding at 3.5 inches a year toward the subduction zone by the Aleutian Islands. Reaching that boundary, the plate dives back into the mantle of the earth, where it is remelted and recirculated. All this—the jungle, the cliffs, the Princeville Resort at Hanalei Bay, the golf courses, the Hawaiian Islands themselves—will not exist in a very short geologic time. All that man and nature have created will be burned and recycled. Perhaps that's where people got the idea of hell. The earth opened up and showed us: there is a fire down below. That's also why the oldest material on this ocean floor dates only to around 100 million years ago. The older rocks have already been recycled.

One evening as I hiked the coast, I found myself on the beach in the shadow of a dark volcanic wall that shot 3,200 feet straight up. I could feel the chill of night caught in the rock. Franklin birds pecked around the shade of a spreading tree, which dropped white, aromatic blossoms. Up on the sand, I saw two figures, white and thin and nude, move together to synchronize.

Here, 21 miles below the Tropic of Cancer, our shadows would disappear beneath us at noon, and now the sun dropped into the ocean, dead-center west with geometric perfection. Unlike in the Arctic, where it can linger for hours on the horizon, the sun was gone quickly here—but the sunset seemed to go on forever. Ranks of clouds formed a convex mantle over the earth, changing color, diminishing to infinity, as they turned to ash around the fading coral coals of daylight. The blue sky waned to gray. The fires went cold. Water turned to lead. The elements of the world were transformed by this subtle alchemy. Ash settled on the mountains as green turned to black, until it looked as if those old volcanoes had smoked up the whole sky. Night came on with startling finality. There was so little ambient light that I could see the glowing plankton of the sea. Then, all at once, the night exploded with stars. I could see the faint mist of the galaxy.

I stood at the water's edge, listening to the surge and cream of the warm ocean as it slowly dissolved the cliffs. The earth takes a big breath of carbon dioxide in the summer as the leaves grow green, and in the winter it exhales it as they fall and die. Canada inhales while Australia exhales, and then the process is reversed. The land churns just like the sea, but in its own time. If you could take one frame of film every five hundred years until you had a two-hour movie, and if you then ran it at the normal speed of twenty-four frames a second, you'd see the whole Hawaiian archipelago undulate like a dragon, breaking the surface with its back in a dozen places like a sea serpent as it surged through time, heading northwest toward the vault of fire and its own destruction in the arms of the earth. Now and then flames would shoot skyward through a vent hole in the ocean floor, and smoke would punch through the planet's atmosphere and circle round, as new islands drifted away and old ones melted down to slag and vanished into the earth from which they came. The entire span of Chinese civilization would be on screen for less than half a second, while Carthage, Rome, and the great European civilizations that followed wouldn't be visible at all.

I cast myself across the land in search of enlightenment, and here is what I found: that matter and energy are one continuous flow. Nothing remains except the process. And that matter is so full of energy that it sometimes has to get right up and dance. And when it does, we call it life.

EPILOGUE

When my father went to war, he and my mother were engaged to be married. He flew a B–17 bomber and was shot down and lost behind enemy lines in Germany in 1945. His parents, Agustín and Rosa, received a telegram from the War Department. (This was in the days before Orwellian euphemisms had undermined the language, and war was not yet called "defense.") The message said that he was missing and presumed dead. Long months of waiting followed, and my mother passed some of the time with my father's sister, Mary Alice, at the little house on Perez Street in San Antonio, where the grieving would occur once the word came at last—probably with the end of the war, while others would be celebrating.

My grandfather—Gus, as he was called—sat in the evenings and on Sunday afternoons by the big old Philco radio, a solid block of mahogany with Bakelite buttons and knobs to tune the stations, and listened to the prizefights and any word that might come from overseas. Winter turned to spring, and Rosa's garden exploded in flower. The chilis climbed the fence, and small green tangerines appeared over the doorway to the back porch, where she sat and painted pottery in the failing light. Roses came out along the front walk, while oleander, gardenia, honeysuckle, and onions blossomed along the driveway that led to the collapsing *cuartito*, a frame garage and storage shed, peeling whitewash back by the alley.

One day in the middle of the afternoon, when everything was quiet, when you could hear the wasps outside the window screens and feel the heat coming up off the earth, the radio was murmuring in a far room. My aunt Mary Alice heard Agustín begin to shout. Then she saw him lumbering through the house, carrying the radio, which was nearly as big as he was. She followed him into the yard, where he ran an extension cord and plugged the radio in and turned it on, leaping and cavorting across the grass. By then the news had ended, and music had begun to play. He grabbed Mary Alice and began dancing with her to the music, as she asked, "What? What, Daddy? What is it?"

"They found him," he said. "They found my boy."

Sonny, they called him. Their pride and joy. Mary Alice was his only sibling. But Sonny was king of the neighborhood, riding his bicycle with no hands, showing off for the girls who watched from their yards, cheering him on, until he ran into the back of a parked car and had the wind knocked out of him and his knees skinned up. Sonny: always looking for the next thrill, the next new thing, the secret knowledge no one knew.

My brothers and I were raised to believe that my father had not only won World War Two, possibly by himself, but had discovered the secrets of the universe, which he held for safekeeping in his lab at the medical school. As a child, I can remember my mother warning us away from him when we'd want him to do magic tricks or wrest miracles out of thin air. "But Mom," we'd protest. "He's not *doing* anything."

"He's thinking," she'd say.

"He's asleep on the couch!" we'd object.

"Well, it's hard work, thinking. Now, come along. Leave him alone." And she'd herd us off and read to us from *Archy and Mehitabel* or *The Wind in the Willows* or a great and mysterious book that purported to contain everything ever known called *The Book of Knowledge*.

In the late 1990s, my father was diagnosed with a very slowly progressing disease of the brain. After subjecting my father to an MRI scan, his neurologist bravely admitted to me, "No one knows what's causing it. Holes are appearing in his brain. They fill with cerebrospinal fluid. Aside from that, there's nothing there. The brain tissue is simply gone."

The symptoms began innocently enough, when my father started having difficulty retrieving common words from memory. I'd come home from one of my journeys and be having dinner with the family, when my father would ask, "So how was your . . . ?" And he'd simply stop, unable to go on.

"Trip?" I'd ask.

"Yes, how was your trip?" And the conversation would continue. Then he began losing his mental maps. He was still driving at the time, and he'd become lost on the familiar two-mile route from his house to mine. He'd show up half an hour late, flustered and apologetic. My mother would say something like, "Well, you know, your father always has a lot on his mind."

The mysterious force continued to eat away at his brain, gradually impairing not only his speech and memory but soon his motor coordination as well. Over a period of six or seven years, he underwent dramatic changes. I can remember sitting with him at the dinner table perhaps as recently as 2000, eating our traditional family meal of enchiladas, beans, and rice, and having spirited discussions about whether life arose on earth *de novo* or was carried here from somewhere else in the universe by wayward comets. His early work in cryogenics left him with no trouble believing that life could travel through the cold vacuum of space. He didn't like the idea of something for nothing, like the utterly baffling complexity and specificity of DNA, when compared with anything inorganic or even compared with the common organic molecules themselves.

At the time of some of our last meaningful conversations, I'd sometimes feel that he and I had come full circle and that I was trying to tell him what he'd been trying to tell me all those years ago. But my explanations left him groping for answers. And he couldn't quite seem to remember enough of the technical details to put together a hypothesis anymore.

In the spring of 2005, I came home from exploring cave dwellings in France to discover that my father was in the hospital and gravely ill. It seemed obvious that he was dying. By then, he could no longer speak effectively on his own behalf, though he still seemed to know what was going on. If I could guess what he was trying to say, he'd say, "Yes," or nod his head. He'd occasionally come out with fragments of original speech, such as, "You know, I think we ought to . . ." But then his storehouse of words would fail him, and he'd fall into silence once more, an expression of sad amusement on his face, as if to say, I always knew it would come to this. I remember sitting by his bed one day after such an episode in which he'd tried to tell me something and failed. I said, "This really sucks, Dad."

He nodded and croaked, "Yeah," and just looked at me with a hangdog expression—a man who once held a hall crowded with scores of people in his thrall twice a week. After many days of attending him, I found myself one afternoon in the basement of the hospital—so like our medical school origins together—and I couldn't help thinking that he, too, would soon be a specimen in someone's lab.

He defied the silence, though, and grew strong enough to come home from the hospital. But a mammoth piece of the landscape that was the Old Man had crumbled in much the same way that the Na Pali coast had fallen into the sea. One day, I asked him what he used to do, what his work was, and he just gave me a puzzled look and shook his head. "I don't know," he said sadly, seeming to recognize that his failure to remember such an essential part of who he was presented us with a bad sign. Something in him was still sentient enough to reflect upon himself, though. And both he and I could do nothing but watch as his vast and varied store of worldly experience and knowledge simply vanished.

He seemed not to remember the war at all. When I asked him about it, he looked into his lap and shook his head. But then sometimes I'd bring out photographs of his dead crew from the B–17 bomber he was piloting when it was shot down, and he'd stare at it and sit silently for a long time. Then, at last, an expression of astonished outrage would come across his face, and tears would form in his eyes. It was his other memory system, the limbic one, sequestered in the depths like the archaea in the sea, still working away.

Paul Davies, the cosmologist, calculated that each of us contains a single atom from each milligram of anything that lived and died more than 1,000 years ago. If he's right, then a few of the atoms that make up my body and your body once made up a part of the Laetoli woman's body and Lucy's as well, and every *Homo erectus* who ever lived. Those atoms have cycled through millions of organisms on the way to becoming us. Some of our molecules have been in a Chinese apothecary and an Egyptian temple as well. In a very real sense, then, I am all people, and you are, too. But we are also all animals and plants. We live delicately poised in a fragile web. Life is certain, but our position in it is not.

As I watched my father die, it seemed as if the entire universe had passed through him, building structure and organization and meaning, and now it was leaking away into space, leaving no trace to suggest that he had ever been here at all. There would be no fossil of his mind, only the increase in entropy that the vortex of his life had made possible. He'd been a transport channel, a convective storm, that had passed across the sea, carrying its accumulated history, the memory of its own momentum. At times like those, I could see that not even the Old Man was the Old Man anymore.

Over time, he went in and out of the hospital, and each time he returned home, there was less and less of him there. Then, in February 2007, he went in one last time, and I took turns with my mother and brother sitting by his bedside, moistening his lips, and fending off a hospital system that could apparently do anything but stop. I remembered him telling me how much he had liked morphine when he'd been severely injured after his B–17 was shot out from under him, so I made sure that he had plenty of it. His breathing grew peaceful with the dreamy opium in his veins. The morning light angled through the window and crept imperceptibly across the wall to illuminate a Currier and Ives print. And then his breathing simply stopped. I put my hand on his head, still warm, and touched his soft, now indefinite-colored hair.

We sent him back to the medical school lab, where he had wanted to go, and he became a specimen for doing what he'd done so well: teaching. He had always liked the idea of things coming full circle. And so he did.

When my father was robust—was himself—I had a habit of calling him up to ask questions, just like the questions I'd asked him long ago. I'd be writing something and realize that there was a fact I didn't know. I'd get him on the phone and say, "Hey, Dad. How's it going? Good, good. I was just wondering: What's the, um, size of the universe?" He'd think a minute and say that the radius was probably between 50 and 75 billion light-years, but that the so-called horizon is only about 14 billion light-years, because that's the approximate age of the universe—actually 13.7 billion, or so they say—but that space has expanded in that time, yielding the much larger estimated numbers, space and time being flexible quantities. I'd say thanks, and we'd chat for a few minutes, and then we'd hang up.

Losing my father's brain was like losing part of my own mind. His amnesia was, in part, my own amnesia. Over decades, married couples can grow together like that. When one dies, the other may fall into a muddle, because he or she literally loses a big chunk of brain function. People who live in close harmony come to depend on each other's knowledge and memory and mental skills; and just as we can delve into another's emotional system and synchronize with or change what's there, we can also extend our own memory and cognitive abilities into the brains of those who are close to us. Losing that is like having a stroke. A part of me had fallen away into the sea.

At some point in my youth, my father gave me a transcript of some lectures that Richard Feynman had delivered in 1964 at Cornell. In one of them Feynman commented on how lucky we were to be alive "in an age in which we are discovering the fundamental laws of nature," but lamented that "that day will never come again." He worried that "the vigorous philosophy and the careful attention to all these things . . . will gradually disappear," that there would be "a degeneration of ideas, just like the degeneration that the great explorers feel is occurring when tourists begin moving in on a territory."

In my lifetime I have seen much of what Feynman predicted come true. Not that we know all there is to know, but that a great deal is now known collectively, and that whatever is truly new is now discovered only by more and more work, yielding less and less, and through more and more obscure means by people sequestered in their specialties. It is not open for all to see, and it is not easy, even while those very discoveries have made the lives of many, like me, far easier. Someone may have all the music ever written by Bach, Beethoven, and Mozart at his fingertips on an iPod and yet have no notion of—and indeed, not even any curiousity about—how the device works. This state of affairs has encouraged us to avoid thinking too much or very rigorously or clearly. It has, as he put it, led to "a degeneration of ideas." And it has focused our mammalian capacity for caring very narrowly. Rather than spreading it more widely, perhaps to the future beyond our own life spans, we have concentrated it on increasingly insignificant things. In our vacation state of mind, we give little thought to what life will be like fifty years from now for the small children laughing and playing at our feet. It doesn't have to be that way.

As I grew up and into the world where I now live, I saw more and more people who seemed to be missing their own lives while hoping to catch the reruns. The sort of paying attention that children engage in naturally, and which was essential for survival in traditional societies, seemed to be disappearing. Against this tendency I struggled to keep my eyes open, to stay awake for the ride, fighting this induced sleep as if against the effects of a drug. How, I wondered, can we wake up for this amazing journey that is so quickly ended? How can we experience the live performance of our own lives? To be in the moment is the ultimate act of redemption. To live with an unquenchable curiosity that sweeps away our mental models and makes everything new is the ultimate triumph we can experience as humans before inexorable forces pull us apart. And it also seems to offer the hope that we might grow up and out of our ape ancestry and into a state where we can live truly examined lives. A truly examined life would be one that gives a gift to the future. It would create the possibility, if not the certainty, that my grandchildren and yours might live as well as we have lived. As matters stand on the earth today, I cannot be sure that even my children will live that well. And the problem isn't that we don't know how to fix the mess we're in. It's that we don't yet believe that we need to.

Although it's easy to pass through life as if in a waking dream, we can enrich our lives, make ourselves more effective, and sometimes even cast a protective web around ourselves and our children, by a habit of knowing—a craving to know—our world and ourselves and by the simple act of consciously paying attention.

SELECT BIBLIOGRAPHY

Asch, Solomon E. "Effects of Group Pressure Upon Modification and Distortion of Judgment." In *Groups, Leadership, and Men*, ed. Harold Guetzkow. Pittsburgh: Carnegie Press, 1951, 177–90.

Atkins, P. W. *The Second Law*. New York: Scientific American Library, 1984.

———. *Four Laws*. New York: Oxford University Press, 2007.

Aurelius, Marcus. *Meditations*, trans. George Long and ed. Irwin Edman. New York: Walter J. Black, 1945.

Barzun, Jacques. *Science*. New York: Harper & Row, 1964.

Bettencourt, Luis M., José Lobo, Dirk Helbing, Christian Kühnert, and Geoffrey B. West. "Growth, Innovation, Scaling, and the Pace of Life in Cities." *Proceedings of the National Academy of Sciences of the United States of America*, April 16, 2007.

Calvin, William H. *The Throwing Madonna*. New York: McGraw-Hill, 1983.

Cantalupo. Claudio, and William D. Hopkins. "Asymmetric Broca's Area in Great Apes." Nature 414 (2001): 505.

Carroll, Chris. "High-Tech Trash." *National Geographic*, January 2007.

Chang, Iris. *The Rape of Nanking*. New York: Penguin Books, 1998.

Columbia Accident Investigation Board. *Report Volume I*. August 2003.

Davies, Paul. *The Fifth Miracle*. New York: Simon & Schuster, 1998.

Dawkins, Richard. *The Ancestor's Tale*. Boston: Houghton Mifflin, 2004.

———. *The Blind Watchmaker*. New York: W. W. Norton, 1986.

———. *The Selfish Gene*. Oxford: Oxford University Press, 1976.

De Duve, Christian. *Vital Dust*. New York: Basic Books, 1995.

Dennett, Daniel C. *Breaking the Spell*. New York: Viking, 2006.

———. *Darwin's Dangerous Idea*. New York: Simon & Schuster, 1995.

De Waal, Frans, and Frans Lanting. *Bonobo*. Berkeley: University of California Press, 1997.

Diamond, Jared. *Collapse*. New York: Viking, 2005.

———. *Guns, Germs, and Steel*. New York: W. W. Norton, 1999.

———. *The Third Chimpanzee*. New York: HarperCollins, 1992.

Dunbar, Robin. *Grooming, Gossip, and the Evolution of Language*. Cambridge, MA: Harvard University Press, 1996.

Eibl-Eibesfeldt, I. *Human Ethology*. Hawthorne, NY: Aldine de Guyter, 1989.

Eiseley, Loren C. *The Star Thrower*. New York: Times Books, 1978.

———. *The Unexpected Universe*. New York: Harcourt, Brace & World, 1969.

Ekman, Paul. *Telling Lies*. New York: W. W. Norton, 1985.

Erwin, Douglas H. *Extinction*. Princeton: Princeton University Press, 2006.

Feynman, Richard. *The Character of Physical Law*. Cambridge, MA: MIT Press, 1967.

———. *The Pleasure of Finding Things Out*. New York: Basic Books, 1999.

Fouts, R. S., and S. T. Mills. *Next of Kin*. New York: William Morrow, 1997.

Frankl, Viktor E. *Man's Search for Meaning*. New York: Simon & Schuster, 1984.

Gell-Mann, Murray. *The Quark and the Jaguar*. New York: Henry Holt, 1994.

Gleick, James. *Genius*. New York: Random House, 1992.

Goodall, Jane. *In the Shadow of Man*. Boston: Houghton Mifflin, 1971.

Gould, Stephen Jay. "This was a man." *New York Times*, June 26, 1996.

Greene, Brian. *The Elegant Universe*. New York: W. W. Norton, 1999.

Grossman, Dave. *On Combat*. Jonesboro, AR: PPCT Research Publications, 2004.

Grove, Andrew S. *Only the Paranoid Survive*. New York: Time Warner, 1999.

Harris, Judith Rich. *No Two Alike*. New York: W. W. Norton, 2006.

Hatfield, Elaine, John T. Cacioppo, and Richard L. Rapson. *Emotional Contagion*. New York: Cambridge University Press, 1994.

Hauser, Marc D. *Wild Minds*. New York: Henry Holt, 2000.

Hawking, Stephen. *A Brief History of Time*. New York: Bantam Books, 1988.

Herodotus. *The Histories*, trans. Robin Waterfield. Oxford: Oxford University Press, 1998.

Hiltzik, Michael A. *Dealers of Lightning*. New York: HarperBusiness, 1999.

Hoffer, Eric. *The Passionate State of Mind*. New York: Harper & Row, 1955.

———. *The True Believer*. New York: Harper & Row, 1951.

———. *Truth Imagined*. New York: Harper & Row, 1983.

Holland, John H. *Emergence*. New York: Addison-Wesley, 1998.

———. *Hidden Order*. New York: Addison-Wesley, 1995.

Johanson, Donald C. *Lucy, The Beginnings of Humankind*. New York: Warner Books, 1981.

Johanson, Donald C., and Blake Edgar. *From Lucy to Language*. New York: Simon & Schuster, 1996.

Kidder, Tracy. *The Soul of a New Machine*. New York: Little Brown, 1981.

Kunzig, Robert. "20,000 Microbes Under the Sea." *Discover*, March 1, 2004.

Lovins, Amory B., et al. *Winning the Oil Endgame*. Snowmass, CO: Rocky Mountain Institute, 2005.

Maclean, John N. *Fire on the Mountain*. New York: William Morrow, 1999.

Malone, Michael S. *Infinite Loop*. New York: Doubleday, 1999.

Mauboussin, Michael J. *More Than You Know*. New York: Columbia University Press, 2006.

McManus, Chris. *Right Hand Left Hand*. Cambridge, MA: Harvard University Press, 2002.

Miller, John H., and Scott E. Page. *Complex Adaptive Systems*. Princeton: Princeton University Press, 2007.

Morowitz, Harold J. *Entropy and the Magic Flute*. Oxford: Oxford University Press, 1993.

———. *Mayonnaise and the Origin of Life*. New York: Scribner, 1985.

Morowitz, Harold J., and D. Eric Smith. "Energy Flow and the Organization of Life." Santa Fe Institute Working Paper, August 7, 2006.

Morris, Simon Conway. *Life's Solution*. Cambridge, MA: Cambridge University Press, 2003.

National Transportation Safety Board. *In-Flight Separation of Vertical Stabilizer, American Airlines Flight 587, Airbus Industrie A300–605R, N14050, Belle Harbor, New York, November 12, 2001*. Adopted October 26, 2004.

Orwell, George. *1984*. New York: Harcourt Brace Jovanovich, 1949.

Pinker, Steven. *The Blank Slate*. New York: Viking, 2002.

———. *How the Mind Works*. New York: W. W. Norton, 1999.

Plotkin, Henry. *Darwin Machines and the Nature of Knowledge*. Cambridge, MA: Harvard University Press, 1994.

Rosenhan, David L. "On Being Sane in Insane Places." *Science* 179 (January 1973): 250–58.

Santoni, G. J. "The Employment Act of 1946: Some History Notes." Federal Reserve Bank of St. Louis, November, 1986.

Sapolsky, Robert M. *Why Zebras Don't Get Ulcers*. New York: Henry Holt, 1994.

Sawyer, G. J., and Viktor Deak. *The Last Human*. New Haven: Yale University Press, 2007.

Schacter, D. L. *The Seven Sins of Memory*. Boston: Houghton Mifflin, 2001.

Schwartz, Jeffrey M., and Sharon Begley. *The Mind and the Brain*. New York: HarperCollins, 2002.

Shapiro, Robert. "A Simpler Origin for Life." *Scientific American.com*, February 12, 2007.

Smith, D. Eric, and Harold J. Morowitz. "Searching for the Laws of Life," *SFI Bulletin*, Winter, 2004.

———. "Universality in Intermediary Metabolism," *Proceedings of the National Academy of Sciences of the United States of America*, September 7, 2004.

Smith, Douglas K. *Fumbling the Future*. New York: William Morrow, 1988.

Smith, Michael. "Fast-Moving Stars Hint At Black Hole." United Press International, September 7, 1998.

Stross, Randall E. *Steve Jobs and the Next Big Thing*. New York: Atheneum, 1995.

Surowiecki, James. "Fuel For Thought." *New Yorker*, July 23, 2007.

Tajfel, Henri. "Experiments in intergroup discrimination." *Scientific American* 223 (November 1970): 96–102.

Tanner, John. *The Falcon*. New York: Penguin Books, 1994. First published in the United States by G. & C. & H. Carvill, 1830, as *A Narrative of the Captivity and Adventure of John Tanner*.

Turnbull, Colin M. *The Forest People*. New York: Simon & Schuster, 1968.

Waldrop, M. Mitchell. *Complexity*. New York: Simon & Schuster, 1992.

Whitfield, John. *In the Beat of a Heart*. Washington, DC: Joseph Henry Press, 2006.

———. "Survival of the Likeliest?" *PLoS Biology* 5(5): e142. doi: 10.1371/journal.pbio.0050142.

Whitty, Julia. "The Fate of the Ocean." *Mother Jones*, March/April 2006.

Whyte, William Hollingsworth. *City*. New York: Doubleday, 1988.

Wilson, Edward O. *Consilience*. New York: Alfred A. Knopf, 1998.

Wilson, Frank R. *The Hand*. New York: Pantheon Books, 1998.

Winer, Barry. Case report no. 1–0041. Tuck School of Business, Dartmouth College, 1996.

Zimbardo, Philip. *The Lucifer Effect*. New York: Random House, 2007.

ACKNOWLEDGMENTS

It is never quite possible to thank everyone who helped with a book. But I would like to single out my brother Stephen, who helped in too many ways to list.

Thanks to Jonas and Betsy Dovydenas for friendship, airplanes, and hideouts in the mountains; to Eileen and Douglas Feldman for the use of their house on the Outer Banks of North Carolina; to my friend and agent Gail Hochman for being there for me for more than three decades; and to my editor, Starling Lawrence.

I would especially like to thank Bill Miller for finding me and inviting me to speak at his company, Legg Mason Capital Management. There I met his chief investment strategist, Michael Mauboussin, who over the next few years provided relentless suggestions for important source material for this book. Bill and Michael also graciously introduced me to the people at the Santa Fe Institute. Thanks to John Miller, Chris Wood, and Geoffrey West for inviting me to visit there. And thanks to Shannon Larsen for keeping me in the loop. While I was at SFI, Murray Gell-Mann patiently answered my questions, and Cormac McCarthy generously read and commented on the first rough draft of this book. That helped me to completely rethink and rewrite it. Eric Smith gave me a voluminous and meticulous technical commentary that helped to clarify difficult subjects.

Thanks to Nita Congress, Robin Easton, Simon Goldbroch, Michael O'Brien, Michael Mauboussin, John H. Miller, Sarah Opdycke, and Kathleen Ratteree for reading early drafts and making valuable suggestions. Throughout the entire process,

Alix Pitcher contributed detailed editorial commentaries that greatly improved the text. Her clarity of mind helped me to focus many of the larger issues here. Thanks also Daniel Callahan for his skillful use of PowerPoint to create illustrations.

I could not have written this book without tireless help and encouragement from my wife, Debbie. My daughters, Elena and Amelia, have been a constant source of inspiration and wisdom. Their mother, Carolyn Lorence, has shown me over many decades what an enriching journey survival can be. And my son, Jonas, has been a great teacher in the art of being in the moment.

Thanks are inadequate to offer my mother, who taught me—among a few other things—about the beauty and uses of language. But I offer her my thanks nonetheless.

INDEX

Page numbers in *italics* refer to illustrations.

Ables, Ben, 51
Abu Ghraib prison scandal, 83, 231
accidents:
    automobile, 44, 139–40
    aviation, 36–45, 47
    avoidance of, 45, 70, 92
    behavioral scripts and, 23–24, 27, 30, 34–45, 51–53, 55–56, 108–11
    coupling events, 44–45
    falling, 55, 56
    filming of, 43
    human error and, 23–24, 37–45, 70, 91–96
    investigation of, 37, 91–92, 95
    study and reports of, 23–24, 43–44
adrenaline, 20, 58, 59–60
Africa, 47, 193, 205, 223
agriculture, 85, 108, 117, 145, 178
air:
    breathing of, 221
    temperature of, 212–13
Airbus, 35–36, 37, 44
air-conditioning, 202
aircraft-pilot coupling event, 44–45
Air Force, U.S., 39
airline pilots' union, 34
airplanes, 144
    bailing out of, 44–45
    cockpit voice recorders on, 41
    crashing of, 47
    fire on, 43
    mechanical problems of, 36–40, 42–45, 54
    small, 19–21, 36, 38, 44–45, 241, 246–47
    spinning of, 37, 43, 44–45
    see also aviation
air pressure, 194, 213
air traffic control, 20, 38
Alaska, 144, 179
Albert Einstein College of Medicine, 73
alcohol, 122, 124–25, 126, 129
algal blooms, 146, 221
Alsing, Carl, 104
Alvin, 193–94, 198, 214
American Airlines flight 587, 34–40, 41–44, 92*n*, 219
American Museum of Natural History, 118
American Petroleum Institute, 149
amino acids, 237
ammonia, 145
amphibians, 221
Anasazi, 244
Andaman Islands, 47, 53
angiosperms, 203, 223
Animal Haven Ranch, 79–80

animals:
  accumulation of chemicals in, 109
  evolution of, 232
  extinction of, 146
  fossils of, 75
  herding of, 108, 228
  intelligence of, 58, 60, 62, 68, 74, 78
  native habitats of, 81, 218, 252
  prehistoric, 58, 114, 202, 204
  as prey, 65, 66, 75
  status in, 79, 80–81, 217
  studies of, 27, 29, 78–82
  threats represented by, 59, 60, 61, 62, 235–36
  use of tools by, 29, 61, 62, 122–23
  warm-blooded, 203
  see also specific animals
antimatter, 234
Antoninus, Marcus Aurelius, 209
apes, 62, 96, 221*n*
  humans as species of, 58*n*, 60, 63, 64, 82, 85–86, 99, 128, 129, 224
  sign language in, 71, 74
  social systems of, 78–82, 85–86
  use of tools by, 29, 61, 62, 122–23
  see also specific apes
Apollo 13, 93
Apollo program, 92, 93
Appalachian Mountains, 193
Apple, 105
  Macintosh computer of, 29, 107
archaeopteryx, 223
archea, 197–98, 214, 263
archeology, 130, 242–46
Arctic, 115, 178, 179, 191
Arendt, Hannah, 143
Arizona State University, 61
arsenic, 173, 178
art, 71, 74, 118
  wall, 66, 187, 201–2, 244–45
Arthur, Brian, 13
Asch, Solomon, 90–91
Assinneboin Indians, 88
asteroids, 238
Atahuallpa, 110–11
Atkinson, Bill, 29
Atlantic Ocean, 12, 14, 193, 223
atomic bombs, 98
atomic nuclei, 190, 234, 236–37, 238
attention, 15, 22, 26, 47, 50–51, 76
Australopithecus afarensis, 57–65, 67, 76, 83
automobiles, 99, 127, 135
  airbags in, 139–40
  alternative fuels for, 149
  crashes of, 44, 139–40
  driving of, 23, 30, 44
  hybrid, 98
  manufacture of, 105, 148
autotrophs, 197–98
avalanches, 166, 167, 168, 204, 222
aviation, 19–21, 26, 27, 34–45
  commercial, 34–44
  commuter, 36
  early, 11, 44–45
  flight training in, 38–39
  ground crews for, 34, 35
  military and transport, 35, 263
  pilot error in, 36–45
  radio and intercom contact in, 20, 35, 38
  role of experience in, 35–36, 37, 38–39
  role of stress and insecurity in, 34, 37, 39–43
  weather and, 19–21

baccarat, 174–75
bacteria, 146, 197*n*, 205, 220
Ballard, Robert, 192–95, 198, 214, 223, 248, 250, 255
ballistics, 66
Bannister, Roger, 28
bargains, 85, 88, 149
baseball, 87, 142
  pitching in, 61, 62, 68
Batopilas, 115, 125, 130
beaches, 11–14
bees, 125, 248
behavioral scripts, 19–31, 82, 109
  accidents and disasters with, 23–24, 27, 30, 34–45, 51–53, 55–56, 108–11
  automatic running of, 23, 30, 31, 33–34, 36, 44, 59, 111
  changing of, 28–29, 31, 34, 44, 45, 49, 69–70, 75
  creation of, 23, 24, 26, 27, 30, 32, 36, 41–42, 48, 52, 73, 125, 140, 224
  efficiency of, 28, 29–30, 59, 67
  emotional systems and, 22, 33, 46, 59–60, 103
  flexible, 117
  generalities in, basis of, 26, 28–29, 46, 110
  learning and practicing of, 22–23, 27, 31, 34, 35, 36, 44, 84

mental models shared among, 23
new, 44, 45, 49
new information ignored with, 19–21, 27, 28, 31, 33–34, 40, 44, 55, 95–96, 110–11
reason and logic pushed aside in, 33, 40
rewards of, 26, 28, 59, 60
smaller scripts within, 44
spurring of, 54–55, 60
tasks simplified by, 22, 28, 60
transferring of, 23
unawareness of, 45
underlying assumptions in, 46
usefulness outlived by, 31, 48
Beinert, Helmut, 249
belief, 114–15, 226
action begotten by, 29
experience vs., 84
seeing and, 31
Bell, Barbara, 78, 80–81
Bell, Gordon, 104
Bell Labs, 106
Benning, Thomas Edward, 186–88
Berreby, David, 87
Betelgeuse (Alpha Orionis), 238–39
Bible, 84, 89, 231
bicycles, 109–10, 260
big bang, 234, 235, 239, 249
birds, 11, 12, 68, 203, 204, 218, 223, 224
Black Cave, 186–87
black holes, 239–40
blindness, 71
blood, 120, 170
bodies, 192
cremation of, 245
dissection of, 159–60
mummified, 186, 188
Boltzmann, Ludwig, 15, 176
bones, 47, 60, 73
bonobos, 60, 78–82, 84, 86, 94
female dominance of, 79, 81
fighting and mating of, 78–79, 80–81
intelligence of, 58, 78
near extinction of, 81
play of, 119
status among, 80–81
use of tools by, 61
botany, 138–39
Bowlin, Mike, 149
bows and arrows, 47, 85, 114, 116, 117–18, 119–21, 165, 243
brachial plexus nerve, 159–60
Braille, 71
brain, 21–22, 56, 146
adaptation of, 71
Broca's area of, 71
Brodmann's area 44 of, 71
chemicals of, 20, 42, 59–60, 233
classification system of, 26, 29
counting and, 228
diseases of, 261–64
frontal cortex of, 63, 65, 67–68, 75, 82, 124
frontal lobes of, 63, 67–68, 124
generalizing and averaging by, 26, 28
growth and development of, 63–66, 70–75, 85, 118, 122, 123, 124, 203, 224
hippocampus of, 71, 187
language and, 61, 70–75
left hemisphere of, 68, 71, 72, 124
limbic system of, 22, 33, 41–42, 59–60, 68, 82, 111, 124, 224, 263
natural workings of, 26, 29
neocortex of, 63, 123
neural functions of, 22, 60, 70, 71, 171, 233, *235*, 236, 239
as organ of experience, 28, 29
planning and rehearsal by, 65, 67–68, 72, 75
plasticity of, 113
selection of mental models by, 25–27, 29
size of, 47, 60, 63–65, 224
social relationships and, 61, 63–64, 223–24
unconscious conclusions of, 24
use of energy by, 203
visual cortex of, 71
see also mind; thinking
Brewer, Garry D., 92–93
Bristol, University of, 89
British Columbia, 49, 50
bureaucracies, 69, 105, 106
business:
big, 77, 98–113, 147
bureaucracy in, 105, 106
CEOs in, 99, 102–7, 111–12, 149
competition in, 102–4, 106–7, 109–10
culture of, 98–113
financial losses in, 104–5, 106, 109–10, 111–13
hierarchy in, 105, 107
industrial, 99–108
mergers in, 109
monopolies in, 100, 102

business (*continued*)
success in, 98–102
wealth generated by, 98–99, 101, 102, 107
Business Week, 109

calcium sulfate, 189
California, 25–26, 48, 49, 50, 138, 179
California, University of, 62, 71
Cambridge University, 228
Campbell, Bernard, 187
Caravaggio, Michelangelo da, 238
carbon, 30, 204, 210, 237–38, 248, 250
carbon dioxide, 30, 144, 145, 146, 177, 178–79, 196, 197, 214
carbon fiber, 148
Carmosino, Penny, 245
Carnegie Institution, 251
Geophysical Lab at, 249
Carpenter, Kim, 242–44
casinos, 172–73, 174, 205, 215
Castaneda, Carlos, 121
catastrophe:
chance and, 48
entertainment based on, 52, 54
global crisis and, 15, 16, 17
cats, 51, 58, 216
caves, 59, 62, 114, 115, 120, 125
art in, 66, 187, 201–2, 244–45
barometric maze, 184, 191
exploration of, 183–92, 262
getting lost in, 186–88, 191, 200
growth of, 185
living creatures in, 190, 199
three-dimensional, rectilinear, 184, 185
underground features of, 183–86, 187, 188–90
cell phones, 173, 177–78
Central Washington University Chimp and Human Communication Institute, 79–80
Chaffee, Roger, 93
Challenger space shuttle explosion, 92, 93–94
change, 32, 145, 148, 224
necessity of, 146, 206
opportunity for, 183, 205–6, 208, 210, 220
quantitative, 238
Charoenkul, Anukul, 47–48
Chauvet-Pont-d'Arc cave, 202
chemicals, 170, 173, 190
brain, 20, 42, 59–60, 233
inorganic, 196
organic, 181
pollution with, 144–47, 149, 173, 177, 178–79, 204, 206–7, 208–9
in water supply, 145–46, 207
chemistry, 13, 16, 59–60, 98, 176, 195
chemosynthesis, 190
Chicago, University of, 75, 196
children, 70, 140–41, 265–66
care of, 57, 59, 64, 127, 139–41, 223–24, 226, 229
death of, 99, 140, 141, 229
development of, 63, 64, 187
future of, 15, 17, 59, 73
learning of, 22–23, 34, 64, 73
play of, 64, 87–88
pranks of, 141–42
work of, 173
chimpanzees, 78–80, 82
brain and body of, 61, 62, 71, 74
fighting and mating of, 78, 123
human beings compared with, 62, 63, 74, 82, 83
intelligence of, 58, 60, 62, 73, 78
male dominance of, 79
strength of, 123
use of tools by, 29, 61, 62
violence of, 79–80
China, 128, 147, 258
circulatory system, 170, 251, 252
cities, 85, 88, 130, 231, 249
citric acid cycle, 214–15, 237, 249, 250
City (Whyte), 227
civilization, 85
clans, 216–17
killing in, 123, 216
survival of, 123, 217
Clausius, Rudolf, 15
claustrophobia, 184
clay, 248–49
climax shape, 235–36, *235*, 239, 244, 248
Closner, Bill, 51
clothes, 123, 186
clouds, 169, 253
coal, 144, 149, 205, 221
burning of, 15, 16, 146–47, 175
cockroaches, 199, 221–22
Cody, George, 249, 250
coffee, 33–34, 153, 154–55, 211
Collapse (Diamond), 108
Colorado River, 68

Columbia space shuttle explosion, 91–92, 93–96, 104
Columbia University, 65
Columbine High School massacre, 141
comets, 195, 238
Communists, 87, 113, 232
computers, 65, 108, 127, 175
  development of, 29, 101–7
  disposal of, 173
  graphical user interface of, 106
  laptop, 98, 113
  manufacture of, 101–7
  marketing of, 106–7
  memory chips in, 111, 112, 178
  personal, 106
  simulation on, 166*n*, 177
  speed of, 138
  3-D programs of, 61
  "windows" concept and, 29
Conn, Herb, 185
continental drift, 193
copper, 173
Copper Canyon, 115, 124, 130, 184, 188
copy machines, 100–102, 108
corn, 122, 136, 137, 145, 146, 202, 243
Cornell University, 227, 265
cortisone, 59–60
cost, 103, 152, 161
  achievement and, 99, 101, 140–41, 177, 206
  hidden, 206
  universality of, 97, 150
cows, 126, 228
Cox, Dobo, 12–14, 16
Cox, Rachel, 191
Crab Nebula, 244
Cray, Seymour, 29
Credit Suisse First Boston Thought Leader Forum, 227–28
Cretaceous explosion, 203
cryogenics, 261
cults, 230, 231–32
culture, 85
  computer, 106
  consumer, 98, 127–28, 135, 162–65, 207
  corporate, 98–113
  differences of, 87–89
  material, 122
  native, 86, 88
  organizational, 85–96
  technological, 16, 91–96, 98–99, 106, 127
curiosity, 14, 15, 20, 27, 31, 46, 49, 152
Custer State Park, 183
cyanide, 141

dance, 74
danger, 15, 19–21
  alertness to, 47
  avoidance of, 165
  signs of, 91
Darwin Awards, 24
Data General Corporation, 104, 105, 106
Davies, Paul, 195–96, 263
Davis, St. James, 79–80
Davis LaDonna, 79–80
DDT, 109
Deak, Viktor, 64*n*
Dealers of Lightning (Hiltzik), 107*n*
death:
  accidental, 30, 34–44, 53–54, 55, 56, 91–96, 140, 141
  childhood, 99, 140, 141, 229
  rituals of, 118, 186, 188, 245
decision making, 21, 60, 72, 76, 77–78, 88, 113
  emotions and, 76, 82
  mistakes in, 25–27, 30, 68–69, 76, 91–96
de Duve, Christian, 195
Deep Survival (Gonzales), 17, 21
dehydration, 70, 187
Deinococcus radiodurans, 199
Dennett, Daniel, 85, 88, 149
Depression, 97–98, 101, 135
deserts, 241, 253
deuterium, 234
Diamond, Jared, 82, 83, 108, 123
Digital Equipment Corporation, 104
digital revolution, 101–7
Ding, Yaping, 228
dinosaurs, 203, 223
disease, 97, 245, 261–64
division of labor, 221
divorce, 93
DNA, 47, 59, 81, 82, 100, 168, 176, 262
dogs, 21, 216
Dominican Republic, 34, 43
Donald, Merlin, 65
Don Juan (guide), 115–18, 120–22, 124–28, 130
dopamine, 59
Dorado computer, 106
Dovydenas, Jonas, 241–47, 253
drumming, 226–27
Dunbar, Robin, 63*n*, 126

Earth, 14, 178, 235
cooling of, 198, 221
crust of, 49, 50, 192–95, 198, 255–56
formation of landmasses on, 222
rotting of, 180, 182, 196, 199
see also sea; world
earthquakes, 47, 48–50, 69, 207, 222, 236
collision of tectonic plates in, 49, 50
destruction in, 50
frequency of, 50*n*
landslides in, 50, 53
measuring force of, 50, 53
subduction zone of, 49–50
Eastern Airlines, 38
ecosystems, 251–52
destruction of, 206–7, 222–23
efficiency, 98, 209–10
automating of activities for, 27
of behavioral scripts, 28, 29–30, 59, 67
Eibl-Eibesfeldt, Irenäus, 86–87
Eichmann, Karl Adolf, 143
Eichmann in Jerusalem: A Report on the Banality of Evil (Arendt), 143
Eiseley, Loren, 202, 203
Ekman, Paul, 225, 226
electricity, 106, 175
cooling with, 202
generation of, 149, 202
heating with, 202, 206
lighting with, 202, 205, 206
static, 54, 100
see also lightning
electromagnetism, 20, 234
electrons, 170, 180–81, 196, 198, 212, 214, 234, 237, 238, 239*n*, 250
elephants, 58, 75, 114, 147, 202
Elon, Amos, 143
emergent events, 168, 203
Emotional Contagion (Hatfield), 226
emotions, 42, 46, 216, 219, 244
behavioral scripts and, 22, 33, 46, 59–60, 103
brain and, 22, 33, 41–42, 59–60, 67–68, 82, 111, 124, 224, 263
caring and, 225, 226, 229–30, 231
central task of, 46
control of, 82
decision making and, 76, 82
evolution of, 58, 224
groupness and, 91, 224–30, 231, 232
synchrony and, 74, 226–28
transcendence and, 74
empathy, 229
Employment Act of 1946, 134
endoplasmic reticulum, 192
endorphins, 60
energy, 174, 195, 201–5
conservation of, 147, 148
dissipation of, 171, 176
geochemical, 14
geothermal, 214
matter and, 258
mechanical, 158
movement of, 152–53, 158–59, 166, 167, 171, 180, 182, 184, 190, 192, 196, 201, 202, 203, 204, 207, 211, 213, 215, 234–35
sources of, 203, 205, 208, 248
spread of, 157, 190, 196, 211, 234, 235
storage of, 203, 205, 211
transforming of, 148, 180, 201, 202, 205, 220–21
waste of, 147, 175, 202, 208
engineering, 29, 30, 94–95, 104, 107, 111
Enron, 66, 110
entertainment, 74–75, 127
catastrophes and, 52, 54
entropy, 157–59, 172–75, 209
debt of, 190, 196, 197, 205, 211, 213, 248
decrease of, 158, 173, 176, 190, 196, 215
definition of, 157
heat and, 202, 205, 239*n*
increase of, 157, 158, 159, 160, 180, 190, 196, 205, 208, 211, 212, 214
measuring of, 239*n*
production of, 174–76, 182, 185, 201–5, 207, 215, 252
environment, 47, 68, 113
adaptation to, 84
alertness to, 47, 50–51, 52, 68, 143
changes of, 48–49, 50–51, 85, 92, 113
conceptualization of, 49
control of, 251
learning and, 44, 47, 50–51
models and scripts suited to, 121
opportunities and perils in, 85, 108–9, 113
pollution of, 144–47, 149, 173, 177, 178–79, 204, 206–7, 208–9
Erin Brockovich, 151
erosion, 60
Erwin, Douglas, 222
Ethernet, 106

Ethiopia, 60, 64
eucarya, 197*n*
eukaryotes, 220
evil, 129, 143, 207
evolution, 14, 15, 28, 34, 47, 49, 56, 85, 88, 118, 201–4, 217, 220–25
  of brain, 63–66, 70–75, 85, 118, 122, 123, 124, 203, 224
  natural selection and, 251
  regression in, 83–84
  sex and, 216, 220–21
experience:
  behavioral scripts drawn from, 59, 103
  brain as organ of, 28, 29
  knowledge and, 25, 26, 132–33, 138
  lack of, 52
  learning and, 59, 103
  recording of, 90
  recounting of, 86
  understanding and, 132–33
extinction events, 146, 204, 207, 221, 222–23, 245
eye-gazes, 228–29

faces:
  expression on, 72, 224–26, 229
  looking at, 228–29
faith, 138
Farmer, Doyne, 169
Far Western Anthropological Company, 242–43
fear, 12, 15, 68, 186, 227
  emotional response to, 36, 40–43, 44, 235–36
  performance degraded by, 39–43, 44
  physical reaction to, 20, 25, 40–43, 44, 128
  of terrorism, 141–42
Federal Aviation Administration (FAA), 19
fertilizer, 145–46, 203
Feynman, Richard, 265
"fight or flight" response, 42, 236
fire, 60, 93, 204
  cooking with, 202
  death in, 30, 93, 95
  earliest use of, 127
  fighting of, 68–70
  forest, 68–70
  setting of, 75, 127, 202
Fire on the Mountain (Maclean), 69
fish, 145, 162, 221, 242, 247
  contamination of, 146, 206–7
fishing, 121, 206–7
fission, 214
five-jawed cradle grip, 61
flags, 87, 88
flight attendants, 35, 38, 40
food, 79–80, 128, 163, 250
  calories in, 164
  contamination of, 109
  eating of, 123, 138, 152, 206, 254
  foraging for, 84, 85
  growing of, 117, 138–39, 145
  high-energy, 202, 203
  hunting for, 75–76, 103, 109
  preparation of, 88, 136–37, 206
  processed, 164, 165, 240
  shipping of, 164
  supplies of, 97, 108, 109, 117, 122, 145, 162–65, 217
Ford Motors, 105
forests, 47, 53, 78, 115, 161, 175
  logging of, 145, 206
  pine, 178
  rain, 165
  spread of, 221
  see also trees
Fortune, 101
Fossey, Dian, 29
fossils, 75, 84, 245
Fouts, Deborah, 79–80
Fouts, Roger, 79
fractals, 170–71, 184, 186, 252
Franz, Carl, 115–16, 120, 122, 126, 130–32
free will, 208, 210
French-American Mid-Ocean Undersea Study, 193–95
fusion, 214, 236–37, 239
future, 76, 217–18
  of children, 15, 17, 59, 73
  imagining of, 28–29, 73
  influence of, 232

Gaia hypothesis, 251, 252
galaxies, 239–40, 241, 253, 257
Galdikas, Birute, 29
Garden of Eden, 84
gas, 16, 172–73, 197–98, 205
gasoline, 135
Gates, Bill, 107
Gell-Mann, Murray, 13, 180, 182, 192, 198, 199, 234*n*
generalization, 14, 26, 28–29

General Motors (GM), 105
Genesis, Book of, 84
genius, 71, 75, 100, 106, 139
genocide, 83, 85
Geological Survey, U.S., 251
geometry, 190, 251
  fractal, 170, 252
Gestapo Department of Jewish Affairs, 143
G.I. Bill, 97
Glacier National Park, 144
glaciers, 144, 179, 221
global warming, 16, 144, 147, 177, 249
gold, 115
Goldman, Jack, 104, 105
Gonzales, Amelia, 20, 129
Gonzales, Augustin, 135, 259, 260
Gonzales, Elena, 20, 115, 130–31, 132–33
Gonzales, Jonas, *22,* 144, 149
Gonzales, Laurence, 134–37, 151–61, 192, 260–61, 266
Gonzales, Mary Alice, 259, 260
Gonzales, Rosa, 99, 135, 156, 259, 260–61, 264
Gonzales, Federico (Sonny), 31, 39, 99, 114, 129, 135, 137, 151–59, 160–61, 192, 196, 220, 239, 259–65
Goodall, Jane, 29, 123
gorillas, 58*n,* 71, 78–79, 225–26
Gould, Stephen Jay, 127
grammar, 70, 72
Grand Canyon, 55–56
gravity, 13, 65, 152–53, 193, 195, 207, 234–35, 237, 238, 239, 255–56
Greek language, 100, 192
Greeks, ancient, 88
greenhouse gases, 177, 178
Greenland:
  melting ice sheet of, 144
  Norse settlers in, 108–9, 111, 115
Green Revolution, 145
Grissom, Gus, 93
Griswold, Wallace L., 245
Grossman, David, 31
groupness, 88–96, 99, 128–29, 224
  cooperation in, 224, 226–27
  corporate, 101, 106, 109–10, 111
  destruction of, 227, 230–31
  emotions and, 91, 224–30, 231, 232
  family and, 230–31
  faulty conclusions and, 92–96, 102–4
  ostracism and, 229–30
  in service of evil, 128–29
  synchrony and, 73, 226–29, 230, 244
Grove, Andy, 111–13
guard:
  "dropping" of, 47, 76, 127, 137
  staying "on," 49, 113
Gulf of Mexico, 146
guns, 31, 40, 80, 86, 110, 115, 122, 124
Guns, Germs, and Steel (Diamond), 123
gypsum needles, 189–90

Haber, Fritz, 145
Haloid Company, 100
Hand, The (Wilson), 62–63
hands, 61–63, 119
  binding of, 66
  development of, 63, 65, 66, 67, 71, 118, 122, 245
  dominant, 72
  gesturing with, 63, 70, 72–73, 74, 75, 116
  grips of, 61, 62
  handling of tools and weapons with, 61, 62, 63, 65, 66
  intelligence and, 71
  shaking of, 66, 116–17
  symbolic significance of, 66, 116–17
  thumbs and fingers of, 61, 62–63, 73, 80, 245
  versatility of, 47, 66, 67–68, 70, 83, 245
Hardy, Oliver, 23
Harris, Judith Rich, 88–89, 91, 228
Harvard University, 27, 146
  Business School of, 97–99, 101, 127
Hatfield, Elaine, 226
Hauser, Marc D., 27
Hawaiian Islands, 254–58
Hazen, Robert, 249, 250
heart transplants, 98
heat, 174–75
  from Earth and sea bed, 192, 194–95
  entropy and, 202, 205, 239*n*
  production of, 158, 211
helictites, 189
helium, 236–37, 248
Hell's Gate Ridge, 68–69
Hernández, Ramona, 136, 137
Herodotus, 9
Herophilus, 160
hexagons, 248
hexavalent chromium compounds, 151*n,* 173
hiking, 25–26, 119–22, 125–26, 129–30, 213, 242–46, 254, 257

Hill, Lynn, 24
Hillary, Edmund, 28
Hiltzik, Michael, 107*n*
Hiroshima, 204
Hitler, Adolf, 232
Hoffer, Eric, 138–39, 230–31
Holloway, Ralph, 65
Holocaust, 143
Homebrew Computer Club, 107
Hominidae, 57–65, 67–68, 74–76, 128, 202, 221*n*
Homo erectus, 64–65, 73–76, 83, 88, 91, 92, 147, 182, 201, 263
Homo ergaster, 64
Homo heidelbergensis, 64
Homo sapiens, 84, 221*n*
Horrocks, Rod, 183–92, 199–200
horses, 41, 110, 116, 141, 202, 243
Houston Medical Center, 152, 155
Howell, F. Clark, 75
How the Mind Works (Pinker), 71, 73
Hubble, Edwin, 233–34
human beings, 16–17
  chimpanzees compared with, 62, 63, 74, 82, 83
  coordinated movements of, 62–63
  evolutionary regression of, 83–84
  evolution of, 56, 63–66, 74, 84, 85, 201–2
  fossils of, 84, 245
  generosity and kindness in, 58, 84, 86
  large brains of, 47, 63, 70, 83
  primitive, 47, 49
  proliferation of, 76, 85
  social relationships of, 61, 63–64, 78, 85–96, 223–27
  success of, 46, 52, 97, 206
  survival strategy of, 27, 56, 68, 216–17, 220
  tribal legends and memory of, 47, 49, 50, 230
  use of energy by, 202, 204–5
  vestigal behavior of, 56, 84, 88, 91
  violence and genocide of, 58, 83, 84, 85, 88
Hungarian Revolution, 113
hunger, 135, 137, 161, 205
hunter-gatherers, 85, 117
hunting, 75–76, 85, 103, 161, 179
  coordination of, 67, 74–76, 219
  restraint and, 75–76
  weapon use and, 62, 63, 86, 116
hurricanes, 151, 177, 212, *212*
Hutchinson, G. Evelyn, 251
hydrocarbons, 16, 205, 206
hydrogen, 195, 196, 197, 198, 205, 214, 236–37, 248, 250
hydrogen bomb, 237
hydrogen sulfide, 198
hydromagnesite, 189
hyperthermia, 141

IBM, 100, 102, 107, 171
ice, 144, 179, 197
  see also glaciers
illusions, 30, 189
  of control, 46, 47, 52, 53, 114–15, 134, 201
  optical, 21–22, *22*, 55–56, 240, 255
  passing on of, 56
images, 21, 26
  satellite, 98, 104
imagination, 28–29, 73, 98, 119
immune system, 112, 216
Incas, 110–11
India, 147
Industrial Revolution, 127, 144, 145, 146
infinity, 192
influenza, 245
information, 234
  confirmation bias and, 94
  exchange of, 126
  illusions overturned by, 30
  processing of, 21–23, 121
  reinterpretation of, 93
  secret, 126, 187
  use of models and scripts instead of, 24, 29–30, 93
  vestigial, 56, 84, 88, 91
insects, 29, 199, 221–22, 227, 242
Intel, 111–13
intelligence, 73, 180
  animal, 58, 60, 62, 74, 78
  cognition and, 73, 77
  context and, 44
  in hominids, 58, 59–61
  learning process and, 44, 219
  mistakes and, 20, 21, 92, 142, 219, 233, 252
  precision, 71
  survival and, 61, 62, 68
"Intelligent Mistakes: When Smart People Do Stupid Things," 13
internal combustion, 202
interstate highway system, 68, 99, 241
Inuit, 108–9
invention, 119, 135, 148, 149, 173, 202

Iraq War, 129, 148
iron, 9, 117, 130, 238
iron sulfide, 249
Ituri Forest, 29, 240

Japan, 50, 51, 174
Japan Airlines (JAL), 38, 39
Japanese army, 128
Jarawa tribe, 47, 53, 85, 91, 96, 114
Jesus Christ, 230–31
jet engines, 34–44, 202, 204
Jobs, Steve, 107
Jockey's Ridge State Park, 11–15
Johanson, Donald, 64, 75, 85
Jordan, Michael, 70
Juan de Fuca plate, 49
Judges, Book of, 89
Jupiter, 238, 253

Kalahari Desert, 84
Kelvin, William Thomson, Lord, 15, 234
Kennedy Airport, 34, 38
killing, 138
  clan, 123, 216
  contests in, 128–29
  of strangers, 123
  in war, 128, 207
  see also death; weapons
knowledge:
  availability of, 51, 52, 138
  diverse, 14, 31
  experience and, 25, 26, 132–33, 138
  learning and, 32, 150
  of the possible, 223
Kodak, 102
Krebs cycle, 214

Laetoli woman, 57–60, 61, 62, 67–68, 76, 77, 88, 96, 119, 134, 139, 241, 249, 263
LaGuardia Airport, 36, 37, 42
lahars, 54
Land, Edwin, 148
language, 60, 70–75, 178, 219, 224
  body, 72, 75, 225–26
  brain and, 61, 70–75
  evolution of, 13, 70–75
  facial expression and, 72, 75, 224–25
  grammar and, 70, 72
  nouns and verbs in, 70, 72
  role of hands and mouth in, 70–73, 74
  sign, 70, 71–73, 74, 75
  spoken, 70, 72, 73, 124, 244
  transformational power of, 74–75
  written and sung, 74, 244–45
Lao Tzu, 103
lasers, 98, 106
Last Human, The (Sawyer and Deak), 64*n*
Laurel and Hardy, 23
Lay, Ken, 66
lead, 173
Leakey, Louis, 29
Leakey, Mary, 57
learning, 76, 152
  accumulation of, 47
  of behavioral scripts, 22–23, 27, 31, 34, 35, 36, 44, 84
  childhood, 22–23, 34, 64, 73
  definition of, 32
  desire for, 14
  environment and, 44, 47, 50–51
  experience and, 59, 103
  intelligence and, 44, 219
  knowledge and, 32, 150
  myopic systems of, 27
  about ourselves and ancestors, 58
  procedural, 27, 121
  questioning and, 159, 160–61
  revision of, 27, 28–29, 31, 68
  of skills, 67–68
life, 167–68
  boundary between death and, 20
  changing ways of looking at, 16, 17, 46
  fragility of, 209
  inside-outside dichotomy in, 215–17, 220, 224, 250, 251
  mystery of, 160
  origins of, 13, 17, 180, 190, 195, 196, 199, 201, 215, 217, 249–50, 261
  return of, 199, 222
  rules of, 14–15, 16, 17
  as self-organizing process, 168
  undersea, 194–95, 198, 221
  unifying patterns and forces of, 17
lightning, 21, 51, 54, 67, 68, 196, 211–12
limestone, 189
lithium, 234
Liverpool, University of, 63*n*
Lotka, Alfred, 176
love, 58, 224
Lovelock, James, 251
Lovins, Amory, 148, 206
Lucy, 60–65, 67–68, 70, 76, 85, 96, 119, 120,

134, 137, 138, 145, 174, 177, 182, 201, 249, 255
brain and body of, 61–63, 65, 263
intelligence and abilities of, 60, 61, 62, 63, 65, 77, 139
Luddites, 144

M-74 galaxy, 172, *172*, 175, 176
MacArthur Foundation fellowships, 146, 148
McCarthy, Cormac, 13, 138, 180
McCologh, C. Peter, 99–106
McGuire, Chloe, 242
McGuire, Kelly, 242
Mackey, Don, 68–70
Maclean, John, 69
magic, 121, 260
magnesium, 238
Makah tribe, 49
mammals, 114, 203–4, 221*n*, 223–26
communication by, 224–27
mimicry of, 225–26
Mandelbrot, Benoît B., 170, 171
manic-depression, 89
map making, 71, 184, 187
marketing, 98, 100–101, 102, 106–7, 163–64, 177–78
Mars, 195, 248, 252, 253
Marzke, Mary, 61
Massachusetts Institute of Technology (MIT), Center for Cognitive Neuroscience at, 71
mass production, 100, 221
mathematics, 15, 161, 207, 215
medical care, 134
memory, 192
consolidation of, 73
creation of, 113, 187
loss of, 28, 261–64
momentum as, 167
of patterns, 74
systems, 152, 167, 168
tribal, 47, 49, 50, 230
Mendenhal glacier, 179
mental health, 82, 83, 89–90
mental models, 19–20, 21–30, 42, 82, 84, 224
behavioral scripts shared among, 23
changing of, 28–29, 31, 49, 50–51, 74, 94, 112–13
faulty, 25–27, 30, 55, 102–7, 109–11
flexible, 117
formation of, 21, 24, 27, 48, 50–51, 52, 67, 73, 74, 112, 140, 187
ignoring information not consistent with, 19–21, 28, 30, 51–53, 79, 93, 94
investment in, 52
new, 49, 50–51, 183
outmoded, 48, 129
possibilities both shaped and constrained by, 29
stability of, 50, 54–55
underlying assumptions about, 46, 54–55, 102, 110
mercury, 173, 206–7
Messner, Reinhold, 28–29
metabolism, 235, 236, 237, 250, 251
metals, 108, 250
meteorites, 195
methane, 197–98
Mexican Revolution, 128, 129
Mexico, 66, 80, 114–32, 227
microbes, 195, 197–98, 217, 250, 251
microprocessors, 106, 111–13
microscopes, 152
electron, 39, 155–56, 189, 220
Microsoft, 107
Mid-Atlantic Ridge, 193
Milky Way, 240, 253
Miller, Stanley L., 196
Milwaukee County Zoo, 78–82, 84, 86, 94, 217, 226
Mimbres, 244
mimicry, 225–26, 228
mind, 152
failures of, 12–15, 16–17, 19–21, 26
freeing of, 20, 22
Paleolithic, 49
"vacation state of," 13, 14–15, 20, 26, 47, 48, 52–53, 77, 114, 135, 147, 149, 201
see also brain; thinking
mindfulness, 31
mitochondria, 192
Molin, Stan, 38–39
Molin, Sten, 34–40, 41–44, 92*n*, 219
money, 174–75, 205, 208
moon, 195, 253
men on, 92, 93
Moore, Gordon, 112–13
Morowitz, Harold, 211*n*, 213, 250
morphine, 264
Mossad, 143
Mount Everest, 28–29, 71

Mount St. Helens eruption of 1980, 51–54, 76, 91
Mozart, Wolfgang Amadeus, 74, 265
multitasking, 30

Narrative of the Captivity and Adventures of John Tanner, A (Tanner), 86
National Aeronautics and Space Administration (NASA), 91–96, 107
  "failure is not an option" slogan of, 93, 94
  fatal accidents of, 91–96, 104
  hierarchy at, 94–95
  organizational culture of, 92–96, 104
  successes of, 92, 93
National Geographic Society, 191
National Museum of Mexican Art, 227*n*
National Museum of Natural History, 222
National Outdoor Leadership School, 191
National Socialism, 232
National Transport Safety Board (NTSB), 37, 41, 43–44
Native Americans, 85–86, 88, 136, 145
natural gas, 202, 204
natural law, 17, 97, 113, 158, 182, 201, 215, 265
  interplay between human behavior and, 16, 87, 202–3
Naval Research Laboratory, 197
Nazis, 113, 142–43, 232
Neanderthals, 123
neon, 238
nervous system, 36, 233, 236, 240
Net-no-kwa, 86
neurons, 22, 60, 71, 171, 233, *235*, 236, 239
neutrinos, 234
neutrons, 234, 237, 238
Newton, Isaac, 34
New York Stock Exchange, 169
"nine dot" puzzle, 55–56, *55*, *56*
1984 (Orwell), 207
nitrogen, 145–46, 248
Nobel laureates, 13, 176, 180, 195
nonequilibrium phenomena, 212–13
nonliving systems, 113, 213
nonverbal communication, 71–72, 224–32
Norman Conquest, 130
North America, 193, 223
North American plate, 49, 193
No Two Alike (Harris), 88–89
nuclear physics, 98, 199
Nuremberg trials, 142

obsidian, 246
Odysseus, 133
oil, 147, 148, 149, 205, 210
  burning of, 15, 16, 175
Ojibwa Indians, 85–86, 88
On Combat (Grossman), 31
Onsager, Lars, 176
opposable thumb, 62
optical illusions, 21–22, *22*, 55–56, 240, 255
orangutans, 58*n*
Orion, 238
Orwell, George, 207, 259
Ottawa Indians, 86
Outer Banks, N.C., 11–14
oxygen, 198, 214, 237, 238
  lack of, 29, 40, 146, 194, 199
oxytocin, 60

Pacific Northwest, 49–50
Pacific Ocean, 12, 14
Pacific plate, 255–56
paleoanthropology, 57, 64, 187
Palevsky, Max, 104
Palo Alto Police Department, 83
Palo Alto Research Center (PARC), 105–7
Pangea, 221
parachutes, 45
paralysis, 40, 116
Paranthropus boisei, 123
particles, 181
  positive vs. negative, 236, 237
  subatomic, 167
patriotism, 88
Pauli exclusion principle, 153, 166
PCBs, 109
Pentagon, 147, 210
People's Guide to Mexico, 115
peptides, 237
permafrost, 197
Permian mass extinction, 222–23
Peru, 110–11
petroglyphs, 243–44, 247
pets, 62, 79
phosphorus, 145, 156
photography, 54–56, 191, 255
  satellite, 98, 104
photosynthesis, 176, 213
physics, 99, 176, 180
  laws of, 16, 190
  Newtonian, 34

nuclear, 98, 199
Pinker, Steven, 71, 72, 84, 124
Pizarro, Francisco, 110–11
plankton, 257
plants, 11–12
flowering, 203–4, 223, 242, 257, 259–60
plate tectonics, *see* tectonic plates
Plotkin, Henry, 28
poetry, 74, 130
police, 229
poliomyelitis, 99
politics, 77, 162
pool balls, 154, 155, 157–58
population, 208
explosive increases of, 145
positive feedback, 228, 230, 237
potassium dichromate, 151*n*
pottery, 132, 244, 260
power law, 207
predators, 47, 67, 177, 228
primates, 57–68
problem solving, 29, 55–56, 123, 152
proprioceptive feedback, 65
proteins, 66, 202, 237
protons, 234, 237, 238
protoplasm, 156
Provincial Emergency Program of British Columbia, 50
puzzles, 55–56, *55*, *56*, 90, *90*
pyroclastic flow, 54
pyruvic acid, 249

Quaker Oats, 109, 240
quantum mechanics, 153, 161, 175, 207, 215, 234
quarks, 234
Queen's University, 228

rabbits, 217–18, 219, 222, 242
radiation, 195, 199
radio, 152, 259, 260
Ramirez, Juan, 116–17, 119, 120, 124
Ramirez, Pedro, 120–21, 129, 188
random arrangements, 175, 176
Ratteree, Kathleen, 64
rattlesnakes, 25–26, 27, 52
Rayleigh-Bénard convection, 248*n*
reproduction, 100
cell, 220
egg-laying, 222, 223
genetic diversity and, 216–17
sexual, 216, 220–21
strategies of, 63–64
reptiles, 12, 93, 203, 204, 222, 225
respiratory system, 170
Richter scale, 50, 53
risk management, 21, 96
Robbers Cave Study, 87–88, 92, 106, 231
rock climbing, 24, 185
Roman Empire, 130
Rosenhan, David L., 89–90
Rubin, Edgar, 21
Rubin vase, 21–22, *22*, 24, 56
running, 28, 71

Sacks, Oliver, 73
Sadiman volcano, 57, 58, 60, 67
safety, 47, 69, 236
assumptions of, 46, 125
quest for, 59, 60, 67, 76, 119, 139–40, 141–42
Salome with the Head of St. John the Baptist (Caravaggio), 238
sand dunes, 11–14
sand piles, 166, 167, 168, 204
Santa Fe Institute (SFI), 13–15, 31, 138, 169, 179–80, 205, 211, 222
Ulam Memorial Lecture Series at, 146–47
Sawyer, G. J., 64*n*
Schacter, Daniel, 27–28
schizophrenia, 89
Scholz, Byran, 69–70
Schrag, Daniel, 146–47, 149, 177
Schwartz, Jeffrey, 71
Schwinn Bicycle Company, 109–10
science, 15, 117, 161
material, 98
medical, 97, 98
Scientific Data Systems (SDS), 104, 106
sea:
chemical pollution of, 206–7
hydrothermal vents in, 194, 196, 198, 217, 249, 257–58
lifeless zones in, 146
living creatures at bottom of, 194–95, 198, 221
rip currents in, 11
rising levels of, 144
surface of Earth under, 192–95
volcanoes in, 193, 194–95, 255–56
Sears, Roebuck and Company, 135

seismic events, 168
self-confidence, 138, 225–26, 228
self-control, 75–76, 123, 124, 132
self-esteem, 230
self-organizing systems, 166–72, 176–78, 184, 190, 203–4, 222, 227
sex:
  animal, 78, 79
  reproduction and, 216, 220–21
  violence and, 78
Shawnee Indians, 85–86
Sherif, Muzafer, 87
Shinto religion, 51
shock, 141, 162
Siemens, 155, 156
singing, 74
Skilling, Jeff, 66
Smith, Eric, 13, 180, 211, 213, 214, 235*n,* 250
Smith, Tilly, 50–51
Smithsonian Institution, 222
social organization:
  of animals, 78–82, 85–86
  brain and, 61, 63–64, 223–24
  cooperation and, 224, 226
  group demarcation in, 82–84, 86–96, 216–17
  hierarchical, 94–95, 105, 107, 110, 217
  hostility and competition in, 87–91, 92, 93
  human, 61, 63–64, 78, 85–86, 223–27
  in-groups and, 89–91, 92–93, 109
social psychology, 82–84, 87–91
sodium, 236, 238
Solid Rocket Booster, 95, 104
South Africa, 118
South America, 194
South Dakota, 183–85, 187–92
space flight, 78, 91–96, 98, 99, 202
  see also National Aeronautics and Space Administration; *specific space missions*
Spirit Lake, 52, 53
spontaneous structures, 166–67
sports, 87–88, 123
sport utility vehicles, 162
Sputnik, 98
Stanford University, 13, 89
  guards-and-prisoners experiment at, 82–84, 87, 231
  School of Medicine at, 62
stars, 236–40, 241, 247
States, Ed, 34–40, 41–43
status, 208, 217
  animal, 79, 80–81, 217
  behavior and, 79, 80–81, 229
  loss of, 66, 230
steel, 105, 110, 148
stock market crashes, 168–69, 222
stock market crash of 1929, 135
stone, 130, 189
  tools of, 62, 136, 243–44
story telling, 74
Straits of Juan de Fuca, 49
stress, 34, 37, 39–43, 128
Strogatz, Steven, 227–28
stroke, 40, 265
stuttering, 72
success, 103
  business, 98–102
  human, 46, 52, 92, 93, 97, 206
sulfur, 195
sulfuric acid, 151*n,* 199
sun, 213, 214, 236–37, 238, 239, 240, 253
  light from, 190, 194, 199, 223, 246
Sunrise Elementary School, 142
supernovas, 238–39, 244, 247
survival, 21, 126, 129, 178, 252
  cooperation and, 224, 226
  creation of new behavioral scripts for, 27, 68
  fight for, 98
  individual, 15
  intelligence and, 61, 62, 67
  strategies for, 27, 56, 67, 216–17, 220
symbolism, 118, 119, 122, 124

Tajfel, Henri, 89
Tanner, John, 85–86, 88, 91, 92, 96
Tanzania, 57–60, 67
Tao Te Ching (Lao Tzu), 103
Tarahumara Indians, 115–16, 117, 120–21, 125–28, 184
Tattersall, Ian, 118–19
Taurus constellation, 244
Taw-ga-we-ninne, 86
technology, 126, 137, 144
  advance of, 137, 147
  computer, 29, 101–7
  culture of, 16, 91–96, 98–99, 106, 127
  failures of, 91–96
  innovation in, 98–108
  modern, 77–78, 86, 91–96, 98–99
  origins and evolution of, 13, 16, 61, 85

practical, 98, 107
primitive, 47, 85
see also specific technical forms
tectonic plates, 49, 50, 192–93, 223, 255–56
television, 98, 136, 173
temperature, 234
air, 212–13
change of, 153, 154–55
cooling, 198, 221, 234
see also global warming; heat
Ten Commandments, 75
terrorism, 141–42
Tesler, Larry, 107
Texaco, 149
Texas A&M University, 30, 95, 96
Thailand, 47–49, 50–51, 52, 69
thermodynamics, 161, 207, 215
second law of, 153, 157, 160, 171, 172–73, 175–76, 182, 189, 202, 205, 211, 234
thermonuclear reaction, 236–37
thinking:
deliberation and reasoning in, 59, 60, 75–76, 114, 210, 219
high-level, 22, 31, 68
in images, 59, 61
intuitive, 69, 70, 76
linear, 55
logical, stepwise, 63, 124
natural, 139
"outside the box," 55
rational, 124, 149
reflective, 123
sequential processing mode of, 67, 69, 70, 72, 73–74, 76, 124
simultaneous processing mode of, 68–70, 76
stream of consciousness, 73
suspension of, 142, 143, 147
see also brain; mind
"three-jaw chuck," 61
throwing, 61–63, 65–66, 67–68, 128
accuracy in, 123
overhand, 61–62
thunderstorms, 20–21, 51, 67, 202, 212, 213
time, 264
idle, 74
shifting of, 217–18
tinkering, 72, 118
Titanic, 192
tools, 63, 73, 204
animal use of, 29, 61, 62, 122–23
deterioration of, 108, 122
hunting, 62, 65, 66
making of, 62, 66, 67, 70, 72, 73, 75, 118–19, 219
stone, 62, 136, 243–44
see also weapons
tornados, 213
torture, 83, 84
touch, 65*n*, 71, 192
Toutle River, 53
trade, 86, 108
training scars, 40, 45
tranquilizers, 89–90
transistors, 152
transportation systems, 209–10
transport channels, 211–14, 250, 263
trash, 127, 162–63
travel, 73–74, 114–15
Traven, B., 115
Treasure of the Sierra Madre, The (Traven), 115
trees, 108, 169, 171, 222
see also forests
Triassic period, 223
trouble, 12, 15, 17, 77, 124
ignoring evidence of, 92, 94–96
Truman, Harry, 134
tsunami of 1700 (Pacific Ocean), 50
tsunami of December 26, 2004 (Indian Ocean), 47–49, 50–51, 85
deaths in, 47, 48, 49, 51
survivors of, 47, 48–49, 50–51, 91, 114
videos of, 47–48
tsunamis, 47–51, 256
areas vulnerable to, 49–50
documentation of, 47–48, 50
earthquakes as cause of, 47, 48–50
killer waves of, 48, 49–50, 58–59
likelihood of, 48, 50
retreat of the sea in, 47, 49, 51, 69
Tufts University Center for Cognitive Studies, 85
tug-of-war, 87–88
Turnbull, Colin, 29, 240
tying, 22–24, 27
Tylenol poisoning, 141

ulnar opposition, 62, 123
unintended consequences, 95–108, 139
United Nations, 178

universe, 149–50, 195
  centrality of life in, 195–96
  expansion of, 233–34
  laws of, 17, 151, 166–69, 175, 195, 201
  origins of, 17, 156, 172, 211, 217, 233–37
  space and matter in, 234–40
University College (London), 28
Urey, Harold C., 196
Us and Them (Berreby), 87

VAX computer, 106
Vertegaal, Roel, 228
videos, 43, 47–48
Vietnam War, 129
Villa, Pancho, 128
vocal cords, fundamental frequency of, 41
volcanoes, 115
  eruption of, 51–54, 57, 58, 59, 60, 61, 67
  lava and hot ash from, 53–54, 57–59, 60, 67, 131, 193, 194, 255
  undersea, 193, 194–95, 255–56
  see also Mount St. Helens eruption of 1980; Sadiman volcano
vortices, 151–54, 157, 159, 167, 168, 171, 192, 203, 212, *212*, 213, 235, 239, 240
Voss, Richard, 171

Wächtershäuser, Günter, 249, 250
wake turbulence, 36–40, 41–42, 43–44
war, 88, 128–29
  costs of, 209
  groupness in, 128
  killing in, 129, 207
  nuclear, 17, 204
waste:
  disposal of, 162–63, 173–74, 177
  electronic, 173
  of energy, 147, 175, 202, 208
  food, 162–63
  of resources, 145, 146, 147, 207, 208
  toxic, 173, 174
water, 62, 70, 115, 126, 248
  chemical contamination of, 145–46
  flooding, 179
  flowing, 158–59, 167, 239, 251
  freezing of, 156
  heating of, 135, 194–95, 202
  human uses of, 207
  ice melting into, 144, 179
  subterranean, 187
  see also sea; vortices
weapons, 122, 134, 138
  modern, 31, 40, 80, 86, 122
  primitive, 47, 85
  steel, 110
  use of, 59, 62, 63, 65–66, 80, 85
  see also specific weapons
weather, 113, 177, 236
  aviation and, 19–21
  satellite images of, 98
West, Geoffrey, 180, 205
whales, 108, 109
White, Ed, 93
Whitfield, John, 204–5
Whyte, William H., 227
Wilson, Edward O., 146
Wilson, Frank R., 62, 65–66, 70
Wilson, Joe, 101–2, 105
Wind Cave, 183–92, 198, 199–200
Wisconsin, 19–21, 26, 27
Wisconsin, University of, 64, 249
wood, 62, 108
Woods Hole Oceanographic Institute, 193
world, 118–19, 122
  exploration of, 73
  human view of, 58, 178
  illusion of control in, 46, 47, 52, 53, 114–15, 134, 201
  symbolic, 118, 119, 122, 124
  understanding of, 32, 45, 117, 133
  see also Earth; environment; sea
World Trade Center collapse, 141
World War II, 39, 98, 101, 162, 259–60, 263
Wright brothers, 11

Xerox, 29, 99–108

Yale University, 92, 176, 251
yawning, 34, 37, 40
Yugoslavia, 74

Zimbardo, Philip, 82–84
Zweig, George, 324*n*